Living on Acreages

What You Need to Know

MWPS-50

LIVING ON ACREAGES

What You Need to Know

NATALIE CARROLL and DON JONES

MWPS
MidWest Plan Service
A Foundation of Knowledge

Designed by Kathy J. Walker

MidWest Plan Service
122 Davidson Hall
Iowa State University
Ames, Iowa 50011-3080

For additional copies of this publication
and a free catalog of other agricultural publications,
call 1-800-562-3618
or
visit our website at: www.mwps.org

First edition, 2006

ISBN: 0-89373-103-X

Library of Congress

The last number is the print number:
9 8 7 6 5 4 3 2 1

MWPS would like to recognize the following authors for providing support to this project:

Natalie Carroll, Associate Professor
Purdue University
Department of Youth Development and Agricultural Education
West Lafayette, IN

Don D. Jones, P.E., Professor and Extension
Agricultural Engineer
Purdue University
Department of Agricultural & Biological Engineering
West Lafayette, IN

MWPS would like to recognize the following external reviewers for providing support to this project:

Brian J. Holmes, Professor
University of Wisconsin-Madison
Agricultural Engineering Building
Madison, WI

William F. Wilcke, P.E., Professor and Extension Engineer
Regional Coordinator for the North Central Region
Sustainable Agriculture Research and Education Program
University of Minnesota
St. Paul, MN

Thomas F. Scherer
Extension Agricultural Engineer
North Dakota State University
Fargo, ND

Joseph P. Harner, P.E., Professor
Department of Biological and Agricultural Engineering
Kansas State University
Manhattan, KS

MWPS would like to thank the following people for providing support to this project:

Frederick Wm. (Bill) Koenig, P.E., Ames, IA
John Moore, Editor, Ames, IA
Letitia Wetterauer, Illustrator, Alpine, TX

Contents

Living on Acreages

What You Need to Know

1

Country Living, an Overview

Congratulations on your move to the country! The general information presented in this book will help you understand some of the opportunities and responsibilities that you now have that you probably would not have if you were to live in an urban area. The discussion will be general, providing an overview of life on a rural acreage. For specific information that applies to your geographic area, you should always contact local authorities and local expertise The major differences between rural living and urban living have to do with the differences in the landscape and your responsibilities and opportunities. You will see and experience more agriculture and forested areas but will live in a less densely populated area, see fewer buildings, and encounter less traffic. Most likely, you will expend more effort taking care of your home and maintaining your property. While your local county governing board will have jurisdiction over many of the decisions affecting your property, those limitations are often not as strict as those in more urban areas. This first chapter provides a brief introduction to rural living and then gives a chapter-by-chapter overview of this book.

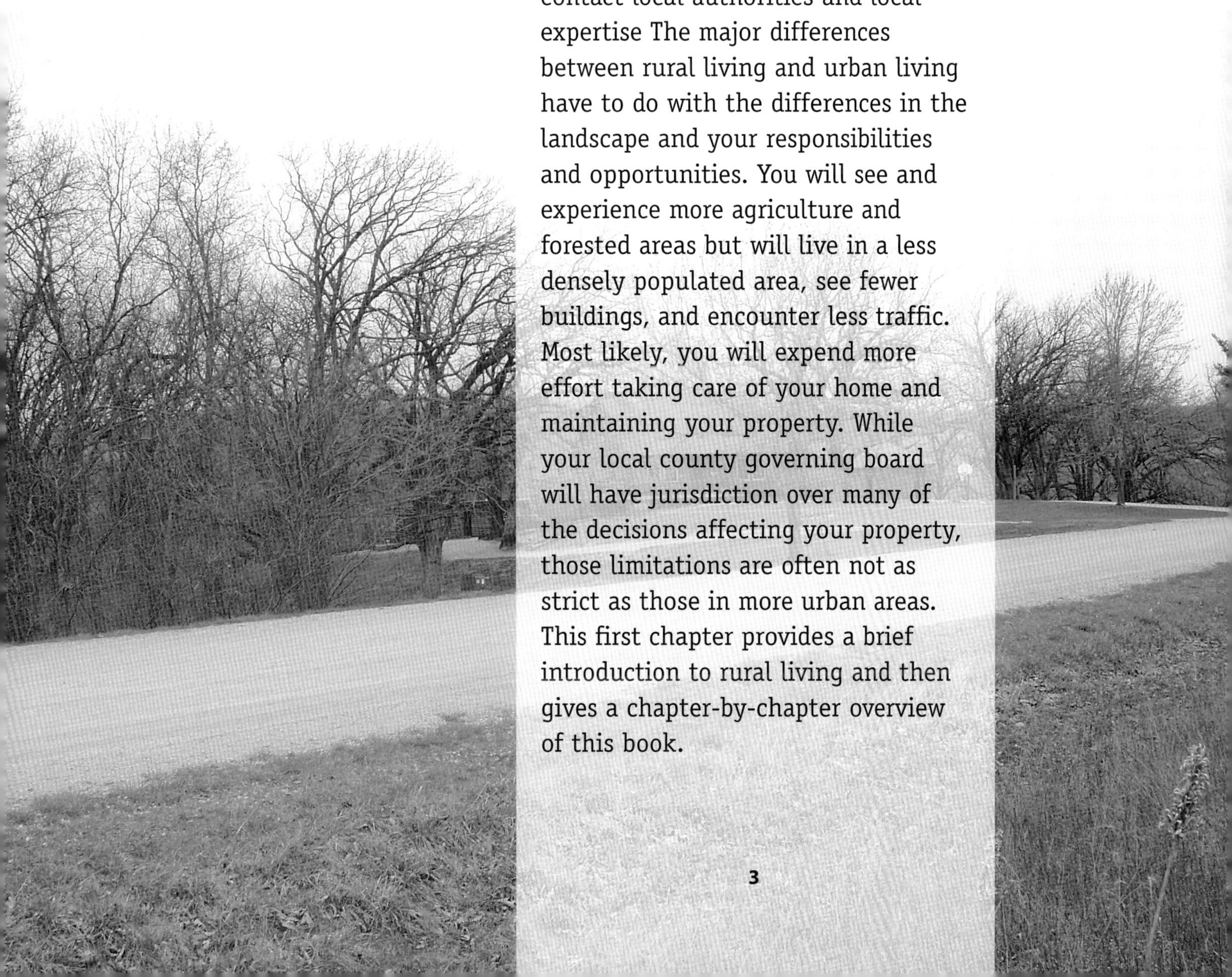

Introduction

The sights, sounds, and odors you will experience in the country will be different from those in the city. Country living is often quieter and more relaxed. Often, you may be able to sit or walk in your backyard without hearing radios, televisions, or other human activities. In the country, people can usually observe more stars in the sky. But the countryside is not all stars and solitude. The open areas and reduced population can create problems that a person moving to a rural area needs to be aware of and be ready to adapt to.

Life in a Rural Community

You will most likely find some differences between urban and rural communities. Although you might be able to enjoy slower and usually quieter daily living, you may also find that a simple trip to the store for bread or milk, or anything else, takes considerably longer. There may be convenience stores near you, but grocery stores, hardware stores, lumber yards, and other major shopping facilities are likely to be considerably farther away. For many who live in the country, it takes a morning (or afternoon) to do their major shopping. If you work in an urban area, you can, of course, tack your shopping on to the end of your day, but the increased distance can make for longer days and increased expenditures for gasoline and other travel-related items.

You may also find the some rural schools are smaller and not as well funded as city schools. This is because the lower population density in the country provides a smaller tax base for funding education. Smaller schools have benefits and drawbacks. Your children probably won't have access to the breadth and depth of courses they would have in a city school. There will probably be less teacher specialization in many high schools, as one teacher may have responsibilities for many topics (e.g., biology, chemistry, and physics). While most rural schools in the U.S. are capable of providing students with a solid education in subject areas that can prepare them for advanced education, it is always advisable to check with neighbors with children to make sure the school will be suitable for your children.

A commonly perceived benefit of smaller schools is smaller class sizes. This doesn't always happen, however, particularly in rural areas experiencing rapid growth. In smaller schools, a smaller student population allows teachers, administrators, and other school personnel to get to know each student and fosters a sense of community in the school. Furthermore, many of the teachers and other school personnel, live in the community and see their students at the local store and at sporting events and other school functions. School personnel may even have their own children in the school system participating in school activities and interacting with your children. It is much more difficult for a student to be anonymous in a rural school. While this lack or anonymity may decrease many types of student problems, it could be problem for parents and students accustomed to larger schools with bigger student populations. The smaller school will give parents more opportunities to be involved in school activities and in their children's educations.

Just as in the schools, rural neighbors generally feel a close sense of community. They may be more likely to notice of who comes and goes than your city neighbors. This may be annoying at first for people who are not used to it, but, after a while, most come to consider the lack of autonomy to be a benefit. Good rural neighbors often keep an eye on your property, as well as their own, and let you know if anything unusual or suspicious is happening in the neighborhood. Depending on your particular relationship with neighbors, you may even want to give them your work number or let them know when you are going on vacation. People sometimes leave house keys with a trusted neighbor in case of emergencies.

Just as in an urban area, you will be responsible for arranging to have your utilities connected. The major utility differences between urban and rural areas are that many people use liquid propane (LP) gas for heating instead of natural gas. You will most likely have your own well and onsite wastewater disposal system (septic system). Both of these systems require monitoring and maintenance for the health and safety of your family. If you get your water from a well, have it checked before you begin using it and yearly thereafter. Also, check your onsite wastewater disposal system periodically, and have the septic tank pumped every three or four years. If you are building a new home, you should review the options for your wastewater disposal system before deciding which one is most appropriate for your site and soils.

You will be responsible for the disposal of your trash. Some areas allow burning of wood and paper products; others do not. You may also wish to compost your kitchen scraps, particularly if you plan to garden because compost is an excellent source of organic matter for flowers and vegetables. Although you may be able to recycle much of your trash, you will also need to find a private contractor to collect your trash. One of the best ways to find out the best way to handle trash or utilities in your area is to ask for advice from your neighbors.

Living with Farming

Fewer than 2 percent of Americans farm for a living today, and only 10 percent of Americans now live in rural areas, but many rural areas, especially in the Midwest, are still heavily farmed. Over time, farms have consolidated and fields are often located several miles apart, and the business of farming continues to require that large and often slow-moving equipment travel on rural roads. Agricultural practices such as planting, spraying, and harvesting, require moving large equipment from the farmstead to cropland. This equipment is not designed to travel at high speeds and may take up most of the road width when traveling from field to field. During harvest, you may also notice large wagons, trailers, or semi-trucks transporting grain to storage or market. In some areas, you may even find Amish farmers who work and travel using horse drawn equipment (Figure 1-1). All vehicles, agricultural and non-agricultural, moving less than 25 miles per hour on maintained roadways must display a slow moving vehicle (SMV) sign on the back of the vehicle that is visible from 500 feet. Be patient and drive slowly and carefully around equipment. Farmers sometimes attempt to pull over for you to pass when it is safe to do so, but if they are driving or pulling very large equipment, it is often dangerous for them to do so. At times like this, it is best to remember that you probably moved to the country to enjoy the slower pace, so try to relax when you are caught behind a large piece of farm equipment moving at 20 mph.

Figure 1-2 shows a variety of farm scenes. As this figure shows, farmers may raise animals and crops together, or animals and crops alone. When farming involves cropping systems, you can expect noise from tractors, combines, and other heavy equipment at certain times of the year. You will especially notice the noise and traffic on the roads during planting in the spring and spraying for pests at times throughout the summer. In the fall, there often will be sounds and dust associated with harvesting. These activities must be done at very specific times and when the weather is right, so you will likely see farmers in their fields around the clock.

Figure 1-1. Amish buggy with SMV.

Cattle resting on pasture.

Field tillage.

Tractor and field tillage machinery traveling between fields.

Grain combine traveling between fields.

Shed and cattle lot operation.

Figure 1-2. Typical farm scenes.

Disruptions from animal agriculture (cattle, swine, dairy) are more consistent throughout the year. When farming involves animal production, you can expect to share the road with slow moving farm machinery and large trucks transporting feed and animals. When animals are being raised in your vicinity, you may be exposed to manure handling equipment, land application of manure, and the odors associated with farm animals and manure application. Because animal odors may be worse during some weather conditions (such as when the wind is from a particular direction) and activities (such as spreading manure), you may notice offensive smells at certain times of the year.

Book Overview

The remaining chapters of this book provide specific information about various aspects of country living. The last page of each chapter provides a list of resources that can provide additional help or more in-depth information for the topics discussed in the chapter.

Chapter 2, *Lot Development*, describes general considerations for new buildings, access to public roads, maintenance and zoning. Because there are many differences across the country, these guidelines are very general. You must talk with local officials about specific information. Chapter 2 can help you decide what questions to ask. Among other things, it explains soil and topography features to consider when you begin searching for a rural acreage or are contemplating improvements to your homestead. Forethought and planning can save many headaches later.

Chapter 3, *Utilities and Rural Services*, describes common water conservation practices and explains water supply systems and onsite wastewater disposal systems. It also discusses electric and gas systems, alternative power sources, trash pickup, telephone services, and mail delivery. You will not be able to get cable television in the country, but satellite dishes and wireless towers can provide services similar to those you may have had in an urban setting. This chapter also discusses permits that may be required in your area.

If you are building a house or other structure, Chapter 4, *Driveways, Parking, and Site Accessibility*, will be useful. This chapter discusses recommended measurements, surfaces, maintenance practices, multiple entrances, and safety. Other topics include mailboxes and school bus shelters.

Chapter 5, *Home Accessibility*, suggests ways to make your home more easily accessible. The chapter contains diagrams and drawings to illustrate the main concepts of universal home design and home accessibility.

Chapter 6, *Outdoor Living Areas*, gives information about caring for your property and outdoor improvements. It provides suggestions for caring for your lawn, deck, pool, or fountain. You will also find information on planting and maintaining a garden as well as suggestions for planting trees and shrubs and for developing wildlife areas. These activities generally require extra storage space, so information is also included to help you build an outdoor storage shed. Example property layouts are included. The chapter also discusses environmental considerations and security issues that you should consider.

Chapter 7, *Pests*, gives suggestions for reducing or preventing damage from rodents, insects, feral cats and dogs, coyotes, raccoons, possums, eagles, and other animals. These animals cause damage to buildings and landscape plantings and can harm your pets or farm animals. Although wildlife often causes problems in urban areas, you can expect more pest problems in your rural home than you had in the city simply because there is so much more wildlife around.

Chapter 8, *Safety, Security, and Emergency Planning*, discusses emergency preparedness, fire protection, ambulance services, snow drift control and snow removal, power outages, weather emergencies, and lightening protection. The more isolated you are from your neighbors, the more important it is that you be prepared and know what to do in an emergency.

A major difference between urban and rural living is that you are responsible for disposal of your household wastes and wastewater in the country. Chapter 9, *Rural Sanitation and Septic Systems*, explains choices for trash disposal and sewage treatment.

Chapter 10 has a *Maintenance Scheduling Calendar* that includes a schedule for common rural household maintenance tasks and suggestions for the jobs you can do yourself and those best left to professionals.

Resources

Cooperative Extension Service: All universities engage in research and teaching, but the nation's more than 100 land-grant colleges and universities have a third mission—"Extension" or "reaching out," and—along with teaching and research—land-grant universities "extend" their resources to help solve problems of public needs. These programs are largely administered through thousands of county and regional extension offices, which bring land-grant expertise to the most local of levels.

Congress created the extension system nearly a century ago to address exclusively rural, agricultural issues. At that time, more than 50 percent of the U.S. population lived in rural areas, and 30 percent of the workforce was engaged in farming. Extension's engagement with rural America helped make possible the American agricultural revolution, which dramatically increased farm productivity and lowered the cost of food.

Call your local county extension office or go to the following website and click on your state to locate your land-grant university's extension service:

Cooperative State Research, Education, and Extension Service (CSREES), http://www.csrees.usda.gov/qlinks/partners/state_partners.html

MidWest Plan Service (MWPS), http://www.mwps.org

Natural Resources, Agriculture, and Engineering Service (NRAES), http://www.nraes.org

Your local Natural Resources Conservation Service (NRCS) office can also be a valuable resource. Or visit at http://www.nrcs.usda.gov/programs/

2

Lot Development

Whether you are building a home or planning to improve your property, you should understand how the location of your home and the soil and the typography of the property affect your pocketbook and your enjoyment of rural living. This chapter discusses locating your house, assessing location for water drainage facilities and utilities, and securing permits for developing your property. The first parts of the chapter apply primarily to new construction.

Locating a House

A farm home should take advantage of the benefits afforded by rural living: space, privacy, and view, while avoiding the disadvantages of being too near noisy, dusty, or odorous farm operations. Your house should be the first building the public sees when entering your property. Locate the house to help control traffic and visitors; place the home upwind from livestock facilities to minimize odors, dust, and insects carried by prevailing summer winds. Orient the house to accent and conform to natural surroundings, enhance family living, and minimize restrictions.

The first step in lot development is to assess drainage around planned buildings and drives. Can the soils and topography accommodate the improvements you are planning? Are any existing drainage facilities adequate, or do they need upgrading? Once you are satisfied with your current situation, you can decide where, when, and how you will develop your property. Many, possibly competing, concerns will come into play. You must consider your soils, topography, accessibility, and safety. Take time to plan new construction or additions, and consider all your options. Consider where water will drain from and around your home; identify where to drill your well and place your onsite wastewater system. Develop a plan for landscaping your property.

The following list contains many ideas to consider if you are locating a new house and developing the surrounding work, play, and living areas.

- Locate your house where the view from the kitchen includes the drive from the public road and your outbuildings. Make the view from the living room the best your site offers (Figure 2-1); use plantings that frame the desirable view and hide neighboring operations.
- Provide space for outdoor family activities.
- Use fencing or shrubs to hide garbage, trash storage, and fuel tanks.

Provide easy access to the patio from the kitchen for outdoor meal preparation, outdoor eating, relaxation, and other family activities.

- Establish surface drainage away from the house to protect the foundation and the basement and to make the areas around your home more usable. Be careful to divert any water draining from farm areas that may be polluted and odorous.
- Place the outdoor family living center inside the protected zone of a tree windbreak but outside the area of the major snow drop. Provide additional wind protection with screening fences or plantings, but avoid blocking desirable summer breezes and causing snow drifts on walks and drives.

An outdoor family living area should be far enough from the public road to allow space for road improvements, to minimize exposure to

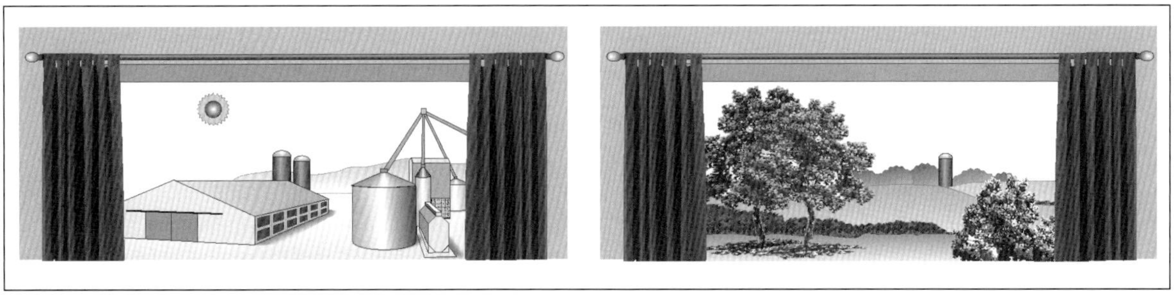

Figure 2-1. Possible views of the farmyard. (Which would you rather look at?)

traffic noise, power lines, and dust and to help discourage vandalism and theft. A 100-foot separation setback from the public road facilitates reasonable year-round access for small to mid-sized farmsteads and also provides room for tree windbreaks, drainage, and driveway maintenance. If future road widening is expected, use a 150-foot minimum. A farmstead with buildings very close to a public road increases child safety concerns and is vulnerable to unwanted traffic and privacy interruptions. The diagram in Figure 2-2 outlines a layout for a family living area.

In addition to proper separation from the access road, locate the family living area at least:

- 100 feet from the shop-machinery area.
- 200 feet from crop storage and processing facilities and from the less-offensive livestock enterprises.
- 300 feet from livestock units with less than about 200 head of cattle or 500 hogs.

Increase these separation distances for larger operations or where major expansion is anticipated. In addition, state or local regulations may dictate that residents maintain a minimum setback from livestock and other facilities.

Ideally, the family living area should be readily visible to those approaching your home. Flowers, shrubs, and landscaping in the family area can help present a pleasing first impression for your home. A strategic location permits the family to observe visitors while maintaining privacy.

Assessing Drainage

Proper drainage around your home is very important. Heavy rain storms and snow melt will cause considerable amounts of water to run off your roof. This water must be moved away from your home quickly to avoid damage to the foundation. Gutters and downspouts must be kept clear.

Heavy rains can also cause erosion in sloped gravel or earthen drives. Careful placement and construction of drives can reduce this effect.

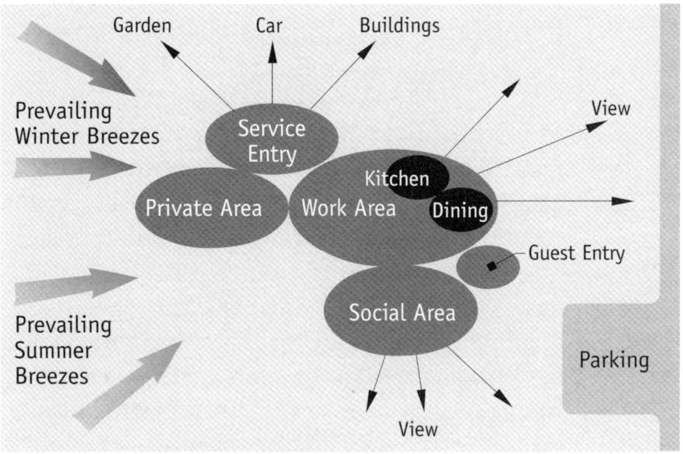

Figure 2-2. Rural family living areas.

To assess drainage, first consider the type of soil on your property. Soils are generally classified by the percentage of sand, silt, and clay they contain. Sandy soils drain much more quickly than loam soils, and clay soils drain more slowly than either sandy or loam soils. The soils around your home will determine the specifications your drainage system. It is very important that your drainage system be properly designed, installed, and maintained, or significant water damage and contamination can result. Sump pumps can be used to remove water occasionally, but they should not be considered a replacement for a properly planned drainage system.

As important as it is, drainage can become one of the most controversial management issues landowners must face. It often entails complex legal and social issues. Local, state, and federal laws can simultaneously apply to a given drainage situation. Consider all levels of jurisdiction before undertaking any actions that may affect drainage. Socially, drainage disputes can strain relations with neighbors. Disagreements can arise between developers, adjacent landowners, and local officials, and disputes can result in court battles. Be sure to consult legal counsel if you become involved in a drainage dispute.

Locating Septic Systems and Other Utilities

Establishing the locations for utilities (septic system, water lines, electric service, gas line, telephone, and other services) is an essential part of planning any modern family living area. The soils around your home will determine the specifications for your sewage disposal system. Having the sewage disposal system on the opposite side of the house from the well or water line minimizes possible cross-contamination. State regulations usually require a minimum of a 50-foot separation distance between the well and the septic system. Check with your local or state health or environmental agencies. Do not locate buried utilities under paved areas unless absolutely necessary. If you must bury utilities under paved areas, place them in conduits and clearly mark their locations on the farmstead layout drawing (Figure 2-3).

If your property has an onsite wastewater system, have it inspected before you purchase the property. If you are unable to do this, contact a professional to have solids removed from the septic tank and your absorption field checked for potential problems. Your local county health department can probably direct you to a reputable contractor. Septic tanks should be pumped every three to five years in most cases, but may need to be pumped more often in some situations.

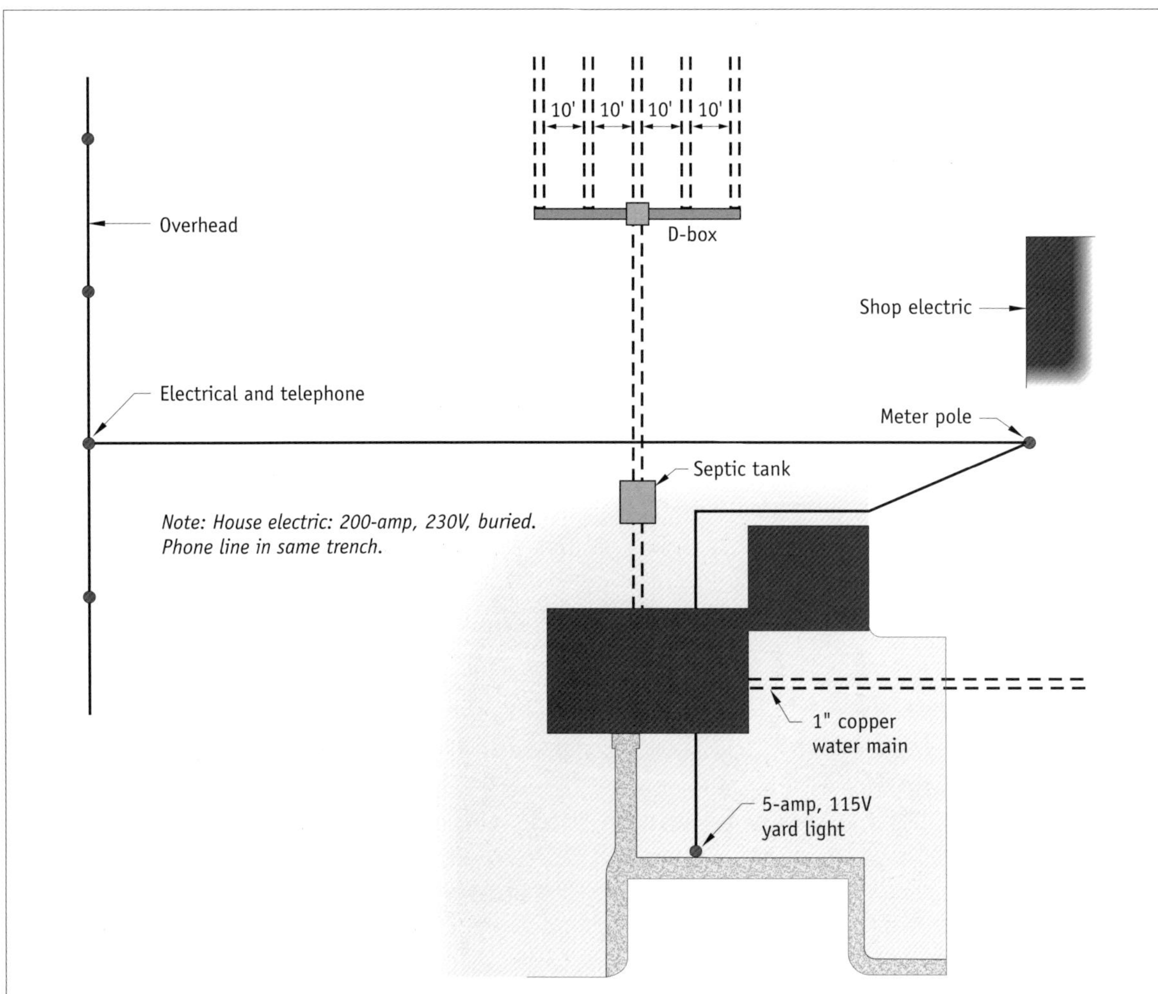

Figure 2-3. Example utilities map for family living area.

Chapter 9, *Rural Sanitation and Septic Systems*, provides an introduction to residential sewage disposal systems. The *Onsite Domestic Sewage Disposal Handbook* (MWPS-24) explains the selection, design, and maintenance of these systems. The *Private Water Systems Handbook* (MWPS-14) contains details about water source selection, water treatment, and water distribution. Consult your county health office about necessary permits and acceptable designs for sewage systems and water supply systems in your area.

The House Handbook (MWPS-16) includes planning for areas inside the home and its surrounding area, and the book *Home and Yard Improvement Handbook* (MWPS-21) explains construction of storage facilities in the house, garage, and yard.

In your planning, consider outdoor family activities. A fenced play area protects small children from vehicle traffic and other hazards. A patio and play area can be shielded from the service yard and driveway by the house, garage, fence, plantings, or a combination of these. Shade is desirable except near vegetable and flower gardens. Deciduous trees provide summer shade but permit sunlight through in winter (Figure 2-4). Evergreen trees shade part of your yard and will block winds year around (Figure 2-5). The effect of trees on wind, snow, and wildlife will be covered later in this book.

Figure 2-4. Deciduous trees shade the house in summer.

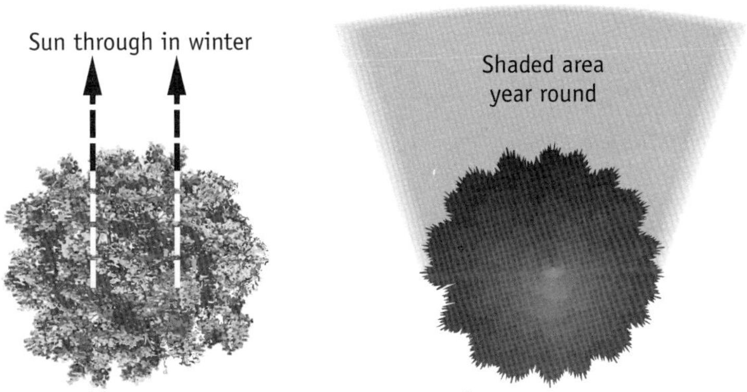

Figure 2-5. Shade from deciduous and evergreen trees.

Obtaining Permits and Inspections

New home construction and many of the changes that you may make over time will require permits and inspections. These are required in most states for activities such as constructing buildings, drilling wells for water, installing septic systems, developing road access (driveways and culverts), and installing oil and gas utilities. Facilities with underground storage tanks containing a regulated petroleum product or hazardous waste substance must register the tanks with the appropriate state environmental regulatory agency. Registration is not required for underground heating oil tanks in all states, so check with the appropriate utility or your state agency.

Before building or renovating, check with your state and local regulatory agencies. Many state and county governments now have websites that explain which permits and inspections are necessary and how to obtain them.

Departments of Fire and Building Services, or similarly named agencies, typically ensure state code uniformity in design, construction standards, and code enforcement for new buildings and building renovations. State environmental agencies typically have the authority to issue several of the state and federal permits required before construction or expansion of various facilities.

County building inspectors typically review plans before construction begins and again after construction is completed to ensure that all construction and safety codes are met. Typically, these inspections focus on electrical and water supply safety. Before any construction, contact your local building officials to check on local requirements. They can explain the type and extent of inspection required in your area.

Working Near Natural Wetland Areas

The U.S. Army Corps of Engineers (COE), the U.S. Environmental Protection Agency (EPA), the state department of Environmental Management, and the state natural resources agency typically regulate wetlands and administer associated permit programs. However, the Natural Resource Conservation Service (NRCS) is the lead agency dealing with wetlands on agricultural land.

The Federal Clean Water Act requires a permit from the COE for dredging and filling in water bodies, including wetlands. The EPA has developed guidelines for COE to use in evaluating dredge spoil disposal sites and has the authority to prohibit use of particular sites. The state reviews all the Corps of Engineers' Section 404 dredge-and-fill applications to ensure the proposed activities do not adversely affect water quality. If wetlands located along streams, rivers, and natural lakes are disturbed, a permit may also be required from your state natural resources agency. For more information, contact:

- U.S Army Corps of Engineers.
- Natural Resource Conservation Service (NRCS).
- Your state environmental and natural resources agencies.

Typically, the state natural resources agency issues permits for construction, excavation, or filling in or on a floodway under the Flood Control Act, which prohibits abodes or residences in or on a floodway. This may also apply for individuals wishing to construct housing alongside streams, even though the home site is above flood stage. Flood insurance may be required before a dwelling can be located in the 100-year flood plain. Check with local authorities for the specific requirements in your area.

Following Other Local and Federal Regulations

If you are thinking of establishing a small or home-based business in a rural area, cities and counties typically have jurisdiction over issues that can affect small businesses. Zoning regulations and signage issues are usually considered to be local issues. Contractor licensing (other than plumbers) is also usually a local option. Some areas also have a "transient merchant" license or some other regulation of those doing business without an established business location. Since regulations vary from city to city and county to county, it is always necessary to contact the local government wherever work is to be done. Consult the local telephone directory for contact information regarding local government offices. For general information regarding the federal government or contact information for any federal agency, contact: Federal Information Center, 800-688-9889.

Resources

MidWest Plan Service, http://www.mwps.org/

- *Private Water Systems Handbook,* MWPS 14
- *The House Handbook,* MWPS-16
- *Home and Yard Improvement Handbook,* MWPS-21
- *Onsite Domestic Sewage Disposal Handbook,* MWPS-24

University of Illinois,

- Drainage Issues on the Rural-Urban Fringe, http://www.extension.uiuc.edu/factsheets/LGIEN%202000-009.pdf
- **Prevent contamination from septic systems,** *57 Ways To Protect Your Home Environment (and Yourself),* ISBN 1-883097-10

3

Utilities and Rural Services

The main difference in utilities between country and urban living is that you are generally responsible for their installation and maintenance. This chapter gives an introduction to electric power, lighting, alternative power supplies, and water supplies and distribution.

You will find it helpful to keep a notebook listing all utility providers: phone, electric, gas, wastewater disposal specialists, trash haulers and others so you can easily find contact information when you need it.

Electrical Power

Power outages in rural areas are typically more common than in urban areas and usually take longer to repair. Homes in rural areas may require standby power, either for convenience or necessity (for example, if someone in the home uses electrically powered medical devices). Typically, electricity is generated at a large, distant power plant and transmitted to the homestead transformer via high-voltage overhead power lines to minimize power loss (Figure 3-1). The farmstead transformer reduces the transmission voltage (usually 7,200 or 14,400 volts) to the single-phase, lower voltage, usually 240 volts, required at the home service entrance. Some large household loads, such as well pumps and air conditioners and ranges, usually require 240-volt service while most of the electrical load requires 120. These circuits are connected to the service entrance box where electricity enters the home. If you have a question about your electric power, consult your rural electric power supplier. A qualified electrician should install or make any alterations to your electrical wiring system.

Typically, wires from the power supplier stop at a transformer located on a yard pole or pedestal at the home lot. Wires from this metered site conduct electricity to buildings and use points around the lot. Conductors from the pole to buildings and the wiring inside building are the responsibility of the homeowner. One meter is usually adequate; however, if a business is located at the site or there are operations such as a large grain handling operation or two homes, separate meters can minimize service conductor sizes and provide information about electricity use for individual enterprises.

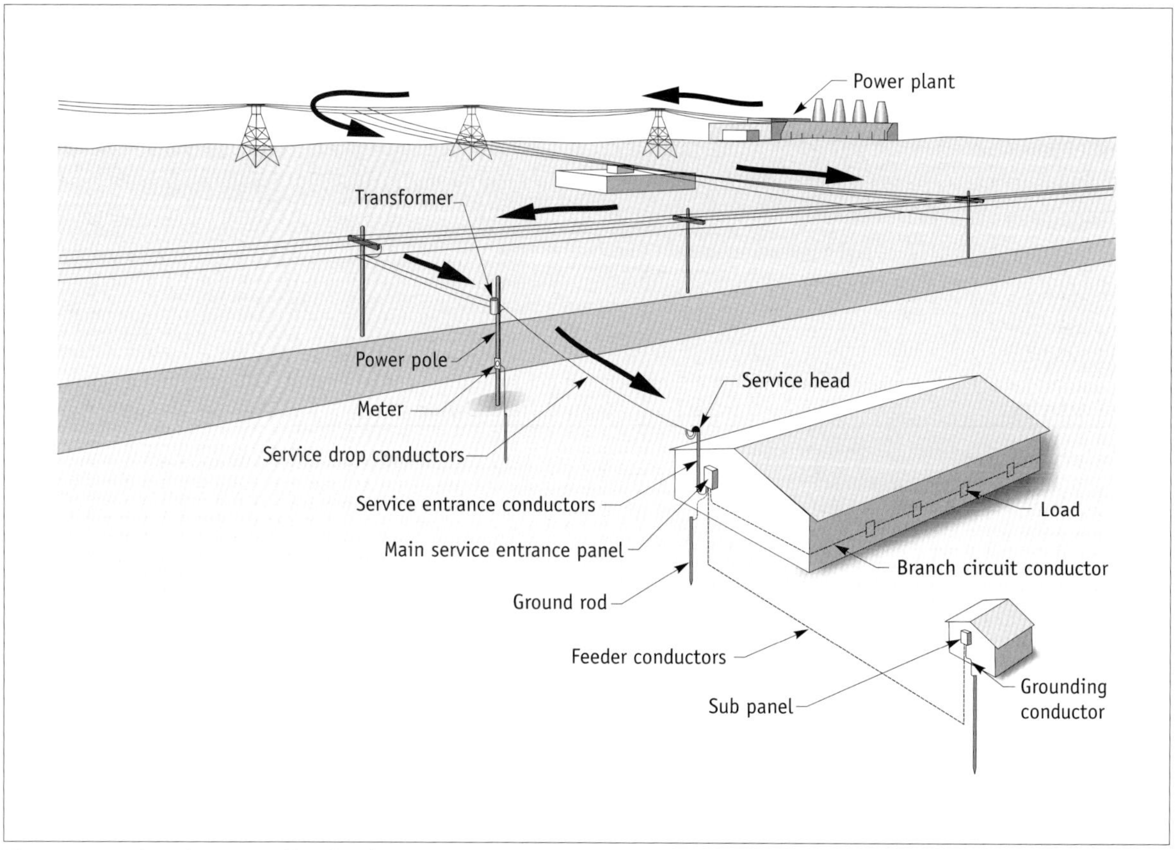

Figure 3-1. Electric power distribution system from the point of generation to the rural home.

Wires from the pole-to-building service entrances may be placed overhead or buried. Overhead wiring allows the use of smaller diameter wire and is simpler to install and modify. Underground wiring requires special wiring and trenching equipment to install but eliminates hazards such as entanglement with tall machinery and service interruptions from ice and wind storms and falling tree branches. Underground wiring is, however, subject to damage from burrowing animals, sharp rocks, corrosion, and excavation accidents. In areas with rocky soil, the use of buried conduits may be justified. Conduits protect the wiring from damage and makes repairs and replacements much simpler. The National Fire Protection Association's (NFPA) National Electric Code sets the minimum requirements for electrical installation. Most states have adopted it, and some have additional requirements. You should keep an electrical layout plan showing the location of all buried electric cables on your property. This plan should identify all buried utility pipes and lines in the area so that you or the contractor can prevent damage from an excavation. **Buried lines should be clearly marked before any construction in the area occurs.** Always contact your local *Call Before You Dig* agency for your area.

Lighting

Exterior lighting promotes general safety and security. When selecting lighting fixtures and locations, take all of the following into account: the size of area to be lighted and the intensity needed for specific tasks; the ease of bulb replacement; problems with bugs, birds, and broken glass contamination; shadows, line-of-sight, glare, and areas such as those near bedroom windows where bright outside light is undesirable. Good lighting is a combination of light quantity, quality, and color. The quantity of light, measured in foot-candles, should provide clear visibility to the work area. The quality of light provided depends on glare, shadow control, and contrast between the light and its background.

Provide an automatically-controlled all-night light near the midpoint in the yard, on a pole or high on a building. Use high efficiency lamps, such as high-pressure sodium or metal halide, and locate fixtures to provide lighting around the home, barns, and adjacent areas. Additional area lights, spots, or floods may be needed on poles or principal buildings for specific areas visited after dark. See Table 3-1 for recommendations. One or more lights on principal outbuildings provide protection against prowlers or predators. Use

Table 3-1. Exterior lighting recommendations.

Source: *Agricultural Wiring Handbook,* 11th ed. 1996, NFEC, Wilmington, OH.

Area or activity	Recommended (foot-candles)	Typical Installation *	Explanation
Security Lighting and Inactive Areas	0.5 to 2	100- to 175-watt hi-pressure for sodium, 25 feet high, 8,000 sq. ft.	Needed ro operator to move about safely.
General work areas and animal feedlots	3 to 5	125- to 250- watt hi-pressure sodium, 25 feet high, for 8,000 sq. ft.	Recommended to discourage prowlers and predatory animals
Active work areas • fuel storage • building entrance • electrical load center • equipment, livestock loading	5 to 10	250- to 400- watt hi-pressure sodium, 25 feet high, for 2,000 sq. ft; supplement with fluorescent on specific areas as needed	Needed for servicing machinery

*Distance between lamps should not exceed 5 times mounting height. Less efficient incandescent lamps of proper wattage may be substituted.

motion detectors and photocells to conserve energy and activate lights only when needed. Check with your local electric power supplier to see if a lease program is available for outdoor security lighting. Yard lights should be on separate circuits. This prevents other circuits from interfering with the control of yard lights and likewise keeps failure of yard light wiring from interrupting service to other outlets.

Sources for Alternative Electric Power

Many rural residents have a gasoline-powered generator to supply electric power in emergency situations. If your electric power goes out because of damage to supply lines in bad weather (not uncommon) a generator can supply a limited amount of power so you can flush toilets, have light, and run your refrigerator, freezer, and other appliances. Generators can also be used to provide supplemental power at distant locations. Other alternatives for distant locations and, possibly, a way to reduce utility costs, are wind and solar electric power generation. Their use is growing and will continue as technological advances improve efficiencies and lower costs. The following alternatives show promise:

- **Ethanol**—Ethanol is a high-octane alcohol produced from corn, wheat, straw, wood or other renewable bio-materials. Ethanol blended with gasoline—usually as a 10% mixture—creates a fuel called gasohol. Ethanol-blended fuels, like gasohol, act as natural antifreeze, and appear to burn more efficiently in combustion engines. Ethanol is non-toxic and water soluble and has the added advantage of being highly biodegradable.
- **Wind Power**—Over the years, wind power technology has progressed significantly with increases in the size and capability of wind turbines. New wind turbines are able to generate approximately ten times more energy than earlier models.
- **Solar Energy**—Solar radiation can be converted into electricity using photovoltaic cells, or can be captured and converted into heat for space and water heating. Solar heating and photo-voltaic systems are both renewable and non-polluting.
- **Hydrogen**—The simplest and most abundant of chemical elements, hydrogen gas can be used as a fuel in specially designed internal combustion engines, or can be used to produce electricity and heat with fuel cells.

Regardless of which system is used, work closely with your local power supplier to make sure that the power generation system is installed correctly. When you are producing electricity, it is critical that your home be isolated from the rest of the power grid. Otherwise someone could get hurt if supply personnel are working on damaged power lines, and your power generator is not isolated from the line.

Electric Load Management

Suppliers of electricity to rural locations, rural electric cooperatives, have load control programs. If a rural electric cooperative supplies power to your property, you are automatically a member of the cooperative and can participate in their programs. Some of the common load management programs they offer are rebates if you buy energy star compliant appliances, install dual heat systems such as ground source heat pumps and electric baseboard heating, and use load control on high-power-use appliances such as water heaters and dual heat systems.

Load control or load management programs are designed to reduce system-wide energy consumption during high use periods by shutting off power for certain uses across the system. By allowing the electric supplier to control the electricity to certain equipment in your house or barn, you can receive a significant reduction in the cost of your energy. If you are building a new house in the country, a visit to the local supplier of electric power prior to construction to discuss load management programs may be very worthwhile.

Water Supply

Rural Americans generally obtain their water from wells and are responsible for both testing and maintaining those wells. However, rural residents in some areas, particularly western states, often are connected to rural water systems that are regulated and tested just like city water systems. Some state and local governments set rules to protect well users. Although the Environmental Protection Agency (EPA) does not monitor private wells, their website (http://www.epa.gov/safewater/privatewells) has information about water quality and some information about wells. The following major well topics are covered at this website:

- The types of drinking water wells and guidelines for proper construction.
- Information about private drinking water wells in your region or state.
- Frequently Asked Questions about your well water.
- Health risks associated with drinking water wells.
- Organizations that work to keep private drinking water wells safe.
- What you can do to keep your drinking water well safe.

This site allows you to download or order copies of brochures, booklets, posters, reports, and multi-media publications. You can also link to web sites with additional information on private drinking water wells.

Water Sources

Drinking water comes from surface water and ground water. Approximately 23 million U.S. citizens rely on private supplies for drinking water. Most of this water is drawn from groundwater through wells, but some households also use water from streams or cisterns. See Figures 3-2 and 3-3 for diagrams of how well supply water. Intermediate storages (Figure 3-3) would be used in the case of a low-output well that is not able to supply water at the rate needed to meet short-term peak needs such as doing laundry or taking showers. If you use water from a stream or cistern, you must take special precautions to protect and maintain your drinking water supply and ensure its quality.

Ground water is pumped from wells that are dug or drilled into aquifers, which are geologic formations that contain water. The quantity of water in an aquifer and amount of water produced by a well depend on the nature of the rock or soil in the aquifer from which the well draws water. Some drinking water wells may be relatively shallow (50 feet or less) while some may be very deep (more than 1,000 feet). Dug wells are generally much shallower (20 to 30 feet) than drilled wells and are not recommended because of the increased likelihood of contamination. However, in some locations, dug wells are the only way to obtain sufficient water, so extra precautions to protect against contamination may be necessary. Groundwater usually does not travel great distances to reach a well, generally less than a mile, so, the activities around your home can have a major impact on your well. For instance, petroleum leaking from a home heating oil tank may enter and float to the top of the water table. Once there, the oil will flow with the groundwater. If your well is in the path of the flow, you may find petroleum in your water supply.

Water use varies throughout the day, and temporary high demands occur each day. A well flow rate of 8 gallons per minute (gpm) is the minimum flow rate to satisfy the water demands in a home, while watering livestock

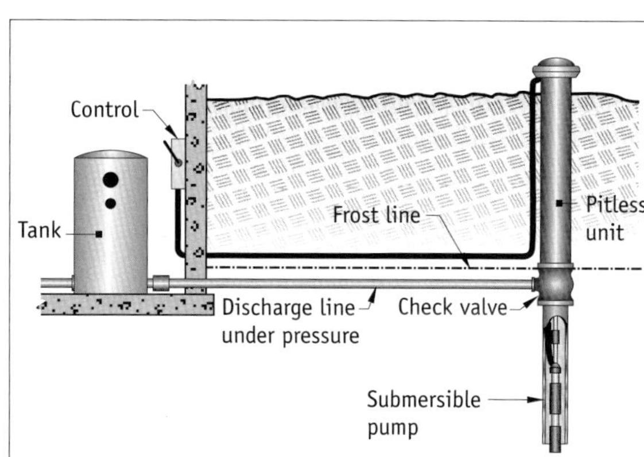

Figure 3-2. Typical well and pressure tank used in many rural homes.

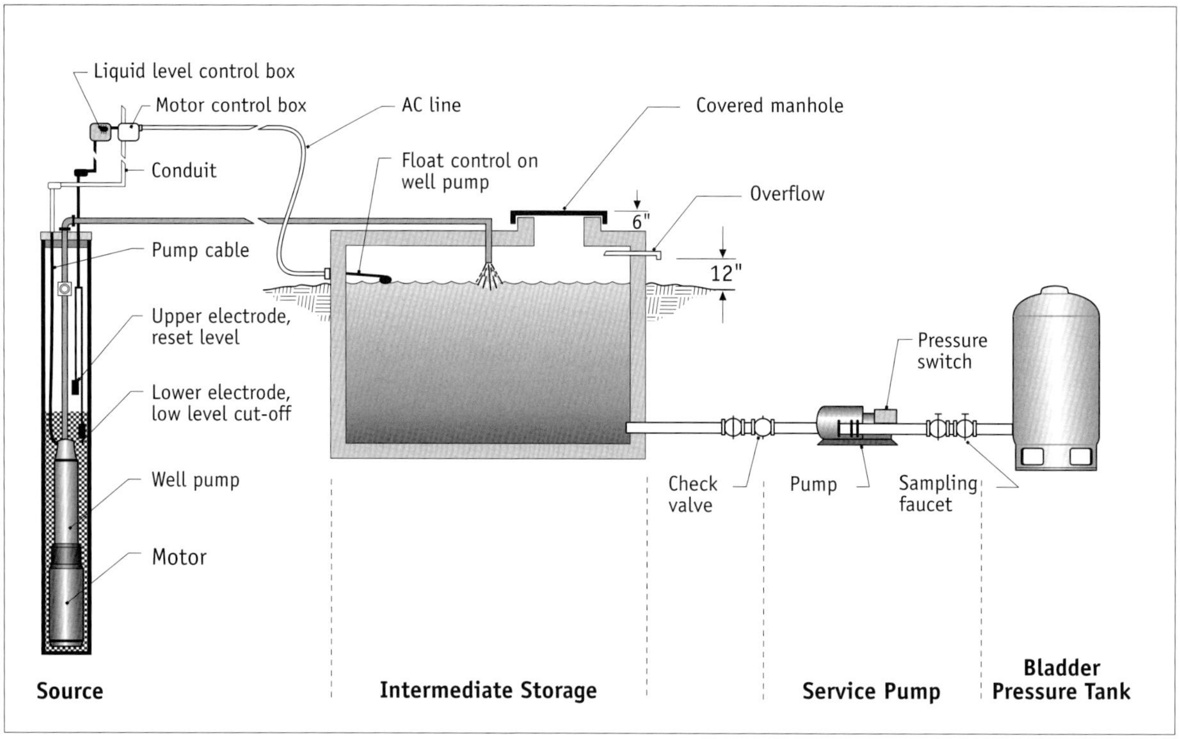

Figure 3-3. Well and intermediate storage with a two-pump system.

and irrigating lawns, fruit orchards, or vegetable gardens add to water requirements. A pressure drop of less than 2 pounds per square inch (psi) is the recommended maximum in the distribution system at a 10-gpm design flow rate. If the well yield is insufficient, it may be possible to install an intermediate storage to meet usage demands (Figure 3-2). There are many ways to conserve water that can result in significant reductions. For example, residential water consumption can be reduced by using water-efficient fixtures (faucets, toilets, and showerheads) and appliances and through better managed lawn watering.

Water Quality

Your family's health, and the value of your home, depends on a safe, high quality water supply. When buying or building a home in a rural area, most people expect high quality water from wells. Yet, that is not always the case. Poor water quality is common. High levels of iron, manganese, and other naturally occurring minerals can make the water undesirable unless it is subjected to costly treatment. Other contaminants, some of which are introduced by human activity and others that occur naturally, can cause significant health problems.

The location of your well is one crucial safety factor to consider. Locating a well properly takes careful planning and consideration of factors such as surface drainage and groundwater flow. A well close to, or downhill from, a road could lead to road salt contamination if the well is not properly constructed. Likewise, locating a well downhill from a livestock yard or a septic system could easily lead to bacterial contamination. In most states, there is a legal minimum distance required between a well and any part of a septic system (usually 50 to 100 feet). In short, attempt to locate your well as far away from potential hazards as practical. This is especially true for shallow wells.

Once a well is installed, be sure to keep the potential hazards such as the following away from your well:

- Storage sheds/garages where petroleum, pesticides, hazardous materials are stored.
- Underground storage tanks.
- Road surfaces (because of the runoff of salt and oil).
- Gardens where pesticides may be used.
- Septic systems.

To help protect well-water quality, do not pour hazardous chemicals down drains or flush them down toilets.

If you plan to drill a new well, consider the costs of pumping equipment, of the energy needed to pump the water, and of any water treatment that may be required. Check with local well drillers, state water regulation agencies, local government offices, and your neighbors for information about local water sources and water well installation requirements that protect water quality against contamination. Be sure to:

- Locate your well on ground higher than potential sources of pollution.
- Build up soil around the well casing to divert water away from the well.
- Avoid areas prone to flooding.
- Make sure the well is properly disinfected after construction.
- Have abandoned wells closed properly.

You can protect your water supply by carefully managing activities near the water source. Several sources of technical assistance are available from your local Extension office (check your local phone book or see the resources section in Chapter 1) to help you protect your water supply. Fact sheets developed as part of the Farm*A*Syst and Home*A*Syst programs help rural residents assess and reduce pollution risks and develop management plans geared toward their circumstances (http://www.wisc.edu/farmasyst).

Water Treatment

It is very important to test your water properly and to do it regularly to assure yourself of its safety. Like almost everything else, a well requires some maintenance. Regular water testing, keeping the well area clean and accessible, and having a driller or pump installer periodically check the well are good practices. Private water supplies should be tested annually for nitrate and coliform bacteria. Every two years, you should test for iron, manganese, chloride, fluoride, pH, hardness, and nitrates. Record and keep the results and observe any changes in the values reported in the test over time. This can help you identify and sometimes fix problems before they become severe.

If you notice changes in the color, taste, or odor of your water, you should have it tested. Such changes may be a sign that your water has become contaminated. Test water more often if someone in the house is frequently ill (particularly stomach distress), and test for lead and nitrate if there is a child or a pregnant woman in the home.

You can choose to install a home water treatment device to add a factor of safety or to address concerns about the taste of your water. Point-of-use (POU) systems treat water at a single tap. Point-of-entry (POE) systems treat water used throughout a house. POU systems can be installed in various places in the home, including on the counter top, at the faucet itself, or under the sink. POE systems are installed where the water line enters the house.

Home filtration units use activated carbon filters, which adsorb organic contaminants and constituents that cause taste and odor problems. Some units can remove chlorination by-products, some cleaning solvents, and pesticides. To maintain the effectiveness of these units carbon canisters must be replaced periodically. Activated carbon filters are not efficient in removing metals such as lead and copper.

Ion exchange units can be used to remove minerals, particularly calcium and magnesium, from your water. They add sodium ions and replace calcium and magnesium ions to soften hard water. Some ion exchange softening units can also remove radium and barium from water. Ion exchange systems that employ activated alumina are used to remove fluoride and arsenate from water. These units must be regenerated with salt regularly to strip the ions removed from the water.

Reverse osmosis treatment units (RO) are more effective at removing contaminants than most other systems. RO units significantly reduce nitrates, sodium, arsenic, and other dissolved inorganic and organic compounds. However, whole-house RO units are not generally installed because for every gallon of purified water they produce, they also produce 4 to 6 gallons of wastewater that goes down the drain. This amount of water can quickly overload a septic system, so a bypass water discharge to a drywell away from the septic systems is necessary. In some states, it is permissible to discharge this wastewater into a ditch or tile since it contains no more pathogens than the drinking water and has extremely low levels of nutrients.

Distillation units boil water and condense the resulting steam to create distilled water. Depending on their design, some of these units may allow vaporized organic contaminants to condense back into the product water, thus minimizing the removal of organics.

Private water supplies should be tested annually for nitrate and coliform bacteria to detect contamination problems early. If a radon or pesticide problem is suspected, the water supply should be tested more frequently.

Many laboratories are available to test water quality. Lists of laboratories certified by your state or the U.S. EPA may be available from your local or state public health department. If you use a private laboratory to conduct the testing, nitrate and bacteria samples will typically cost between $10 and $20 per sample (2005 price). Testing for other contaminants will be more expensive. For example, testing for pesticides or organic chemicals may cost from several hundred to several thousand dollars.

Wellhead and Watershed Protection

Wellhead protection is a process of identifying potential well contamination hazards in the area surrounding a drinking water well (the wellhead). Contamination in the wellhead area could affect your water supply for many years. A community wellhead protection plan includes five components. A wellhead protection plan for the well at your home should include items one through four of this list.

1. A map of the area that has the greatest potential to contaminate the well (the wellhead protection area).
2. An inventory of potential sources of contamination in that area.
3. A management plan to reduce potential contamination from those sources.
4. A contingency plan for emergencies.
5. Public participation, education, and outreach.

Heating

Propane or Liquefied Petroleum (LP) gas, fuel oil, and wood are the fuels most commonly used for home heating.

Fuel Storage and Handling

LP gas is stored either in above-ground or buried cylindrical tanks available in sizes from 50 to more than 2,000 gallons. The most common size of a household storage tank for fuel oil is 265 gallons. The tank can be located either outside next to the house or inside, such as in the basement. The location of a fuel storage tank, whether for heating or vehicles, is safest away from busy traffic ways yet must be accessible for filling in any season. Unlike lighter-than-air natural gas (methane), propane gas is heavier than air and can accumulate in enclosed or low areas and cause an explosion or fire hazard.

Shade and white/shiny paint help lower the temperature of an above-ground LP gas tank in hot weather and add visual enhancement to the tank site for safety. A reinforced concrete pad under a fuel tank prevents uneven settling, eliminates weed growth, and aids neatness. The local fire department, the gas supplier, and your insurer may have preferences about tank location and installation. National Fire Protection Association #58 *LP-Gas Code Handbook* explains criteria for installation, operation, and maintenance of propane storages (Figure 3-4 and Table 3-2).

Table 3-2. Typical separation distances requirements for propane tanks.
Adapted from *Liquefied Petroleum Gas Code,* NFPA58 and *The National Fuel Gas Code,* NFPA54.
Check with your state agency for specific requirements in your area.

Rated Tank Capacity	Minimum Distances between Tank and Building, feet		
(Gallons)	Above-ground	Under-ground	Between above-ground Tanks
Less than 125	None	10	None
125 to 250	10	10	None
251 to 500	10	10	3
501 to 2000	25	10	3
More than 2000	50	50	5

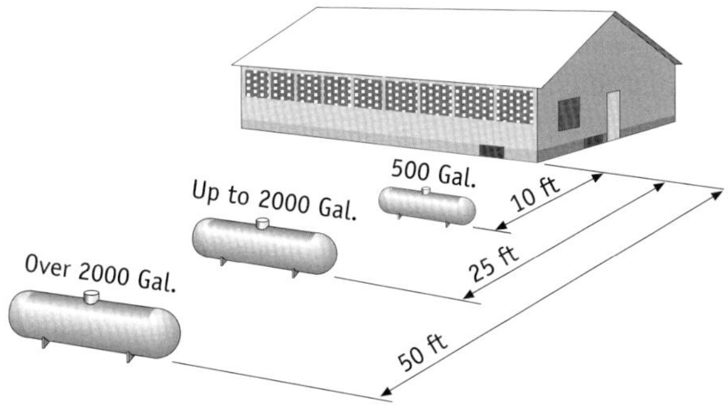

Separation distance required from buildings.

Fuel storage tanks in a concrete containment.

Figure 3-4. Propane (LP gas) storage.
Adapted from Appendix G, *The Liquefied Petroleum Gas Code,* NFPA58.

Alternative Heating Sources

Propane and fuel oil are the most common home heating sources. However, some rural homeowners rely on wood, coal, electricity, or other fuel sources to heat their homes. Biomass fuels (plant-based) are not new. In fact, until about 150 years ago nearly all homes were heated with wood or coal. Newer alternative heat sources include ground source heat pumps that can be very efficient and are able to provide both space heating and cooling. A ground source heat pump system uses a series of buried pipes and a heat pump to transfer heat from the ground or groundwater into a building during winter, where heat is extracted to warm air distributed throughout the home via ducts. The same system can also heat water. In summer, the system is reversed to transfer heat from the building to the cooler ground, which serves as a heat sink.

Biomass refers to fuel derived from plant materials such as wood, agricultural crops, and other organic residues. Wood is the most common biomass fuel. New wood-burning fireplaces are much more efficient than earlier versions. In cold climates where wood is readily available, some rural residents also use wood stoves located outside the home to heat hot water. Hot water is recirculated through the home and back to the stove. These systems are designed to have fuel added only once a day. A major advantage is that the ash, dirt, and dust associated with handling and burning wood remain outside.

New equipment is also becoming available that can use a variety of biomass materials such as corn (both stover and grain), cherry pits, peanut shells, and sunflower seeds for fuel. One 50-pound bag of corn or 40 pounds of wood pellets will heat a modest home for a day or two except in very cold weather. As a rule, one acre of farmland can produce enough corn to heat two average homes located in the U.S. Corn Belt for the entire year. One ton of corn or wood pellets contains enough heat potential to replace three barrels of oil. Several biomass fuel sources (corn stalks, cherry pits, sunflower hulls, switchgrass) are being studied for use regionally.

Safety

Wood stoves, furnaces, fireplaces, (and appliances such as gas clothes dryers, gas stoves, and gas water heaters) are combustion units and, as such, require safety precautions. They increase the demand for combustion oxygen and produce carbon monoxide. If adequate fresh air is not provided, these combustion units pose a danger of carbon monoxide poisoning or suffocation. Efforts to make homes more weather-tight with improved caulking, siding and/or insulation, and better windows and doors, may restrict the amount of combustion air available. Therefore, outside air should be provided using a vent or an air exchanger that warms incoming cold air before it enters the home.

The chimney of an alternate fuel heating system needs to be cleaned and checked each year to prevent chimney fires. Masonry chimneys should be inspected for cracks, crumbling mortar, obstructions, and creosote deposits. Prefabricated metal chimneys need to be inspected for corrosion, tightness of the joints, and creosote deposits. Chimney cleaning and inspection should be done by qualified, trained persons. It is not a job for amateurs or the average home owner. You should also check your wood burning stoves or fireplace each year. Look for cracks and other defects such as leaky door seals, broken or loose hinges, and faulty draft regulators. Examine the legs of wood stoves to make sure they provide a sturdy and secure base for the stove. If a wood stove has a cracked glass insert in the door, it should be replaced.

A slow burning fire creates more creosote than does a faster one. When creosote buildup occurs frequently, adjust the draft of the fire to provide more air and speed up combustion. An occasional faster burning fire can help to reduce creosote buildup.

Be sure that your home is protected with working smoke and carbon monoxide alarms. Some insurance companies provide discounts on insurance premiums to clients who have smoke and carbon dioxide alarms in their homes. Test these alarms regularly and replace the batteries annually, such as on your birthday or a holiday. See Chapter 10 for a sample maintenance schedule. Do not remove batteries because of a false alarm (due to occasional smoke from cooking or the fireplace). Careful location of the detector and better management of the wood stove can reduce false alarms.

Other Services

Services such as telephone, mail delivery, television, and internet in rural areas sometimes differ significantly from such services in urban areas. Anyone moving to a rural acreage should be aware of this and prepared to deal with the disparities.

Telephone Service

While telephone service in rural areas is very similar to that in urban areas, there may be more frequent disruptions caused by storm damage to telephone lines. However, cell phone service in many rural areas is of very good quality because there are more signal towers and fewer large obstructions to block the signal.

Mail Delivery and Mailboxes

Note: The following information is adapted from the United States Postal Service website: http://www.usps.com/

Rural area mailboxes are vulnerable to vandalism because they are usually isolated, located on public thoroughfares, and frequently not visible to the box owners from their homes. Mailboxes are considered federal property, and federal law makes it a crime to vandalize them (or to injure, deface, or destroy any mail deposited in them). Violators can be fined up to $250,000, or imprisoned for up to three years, for each act of vandalism. Theft or destruction of mail or mailboxes should be reported to your postmaster as soon as possible.

The receptacles may be used only for matter bearing postage. No part of a mail receptacle may be used to deliver any material not bearing postage, including items attached to or hung from the mail receptacle. Homeowners must keep the approach to their mailboxes clear of obstructions to allow safe access for delivery. If USPS employees are impeded in reaching a mail receptacle, the postmaster may withdraw delivery service. Although curbside mailboxes are to be used for mail only, publishers of newspapers regularly mailed as periodicals may, on Sundays and national holidays only, place copies of the Sunday or holiday issues in rural route and highway contract route boxes of subscribers if those copies are removed from the boxes before the next scheduled day of mail delivery.

Mailboxes must be made according to government specifications. If you wish to make a custom-built mailbox you will need the approval of your local postmaster. The custom-built box must generally meet the same standards as approved manufactured boxes for flag, size, strength, and quality of construction. Every curbside mailbox should follow these rules:

- A box number, if used, should be inscribed in contrasting color in neat letters and numerals at least 1 inch high on the side of the box visible to the carrier's regular approach, or on the door if boxes are grouped.
- A box should have a house number if street names and house numbers have been assigned by local authorities, and the postmaster authorizes their use as a postal address. If the box is on a different street from the customer's residence, the street name and house number should be inscribed on the box.
- Some homeowners add their name to the mailbox, but this is discouraged by authorities in many areas for security reasons.
- Advertising on a mailbox or its support is prohibited.
- The post or other support for a curbside mailbox should be neat and of adequate strength and size. It may not represent effigies or caricatures that tend to disparage or ridicule any person. The box may be attached to a fixed or movable arm. While the post and box should be sturdy, they should be should not be so strong or massive that they could cause serious injuries in the event of a vehicle crash.
- Subject to state laws and regulations, a curbside mailbox must be placed to allow safe and convenient delivery by carriers without leaving their vehicles. The box must be on the right-hand side of the road in the direction of travel of the carriers on any new rural route or highway contract route, in all cases where traffic conditions are dangerous for the carriers to drive to the left to reach the box, or where their doing so would violate traffic laws and regulations.
- A mailbox with a lock must have a slot that is large enough to accommodate the customer's normal daily mail volume. The USPS neither opens a locked box nor accepts a key for this purpose.
- A receptacle for newspaper delivery by private carriers may be attached to the post of a curbside mailbox used by the USPS if the receptacle:
  - Does not touch the mailbox or use any part of the mailbox for support.
  - Does not interfere with the delivery of mail, obstruct the view of the mailbox flag, or present a hazard to carrier or vehicle.
  - Does not extend beyond the front of the mailbox when the box door is closed.
  - Does not display advertising, except the publication title.

Television and Cell Phone Signals, and Internet Accessibility

Cable television is not an option in rural areas because the low density of homes makes running cables long distances very expensive. Many rural residents depend on satellite dishes and/or television antennas to increase the number of channel choices. Internet access will probably be via phone lines although satellite dishes and combination plans for television and Internet access are gaining popularity. Modem access may be slow, but cable is seldom available, and DSL lines must be located near switching equipment, which is normally in urban areas. Cell phone towers are common in rural areas, but signal strength can vary greatly due to location and terrain.

Resources

Alternative fuels—From "Investing in Manitoba, Business Facts,
http://www.gov.mb.ca/iedm/invest/busfacts/utilities/utility5.html

American Wind Energy Association:
http://www.awea.org
122 C St NW, 4th Fl Washington DC 20001, current developments for use, location, and size selection of wind power equipment

Drainage Issues on the Rural—Urban Fringe, LGIEN Fact Sheet, 2000-009, University of IL Extension

Emergencies: Report oil or chemical spills at (800) 424-8802

EPA website, http://www.epa.gov/
Topics include: Acid Rain, Air, Asbestos, Clean Air Act, Clean Water Act, Cleanup, Drinking Water, Enforcement, Global Warming, Grants, Ground Water, Hazardous Waste, Human Health, Lead, Mercury, Mold, Oil Spills, Ozone, Pesticides, Radon Recycling, Storm Water, Superfund, Surface Water, Technology, Test Methods, Wastes, Wastewater, Water, Water Pollutants, Wetlands, and more.

Farm*A*Syst and Home*A*Syst,
http://www.uwex.edu/farmasyst/index.html

MidWest Plan Service (MWPS),
http://www.mwps.org

- *Onsite Domestic Sewage Disposal Handbook*, MWPS-24
- *Private Water Systems Handbook*, MWPS-14
- *Wiring Handbook for Rural Facilities*, MWPS-28

National Fire Protection Association,
LP-Gas Code Handbook (#58)

National Food and Energy Council (NFEC)
PO Box 309, 2333 Rombach Ave., Wilmington, OH 45177 (http://www.nfec.org):

- *Agricultural Wiring Handbook*
- *Electrical Wiring for Livestock and Poultry Structures*

Natural Resource, Agriculture, and Engineering Service, Cooperative Extension, Ithaca, NY.,
Private Drinking Water Supplies, Quality, Testing, and Options for Problem Waters, NRAES-47

Solar Energy Industries Association,
122 C St NW, 4th Fl Washington DC 20001, current developments for use, location, and size selection of solar power equipment

The National Sanitation Foundation,
http://www.nsf.org, (800) 673-8010

The Water Quality Association,
http://www.wqa.org, (800) 749-0234

US Department of Energy: http://www.eren.doe.gov

USPS—US Postal Service,
http://www.usps.com/

Wind resource map,
http://www.nrel.gov/wind/usmaps.html

4

Driveways, Parking, and Site Accessibility

This chapter describes considerations for residential driveways, parking, culverts, yard lights, and guest parking in a rural area. Careful planning and maintenance will assure safe access. See Chapter 5, *Home Accessibility,* for driveway and sidewalk accessibility considerations.

Homeowners typically select the location of their house very carefully to achieve specific views of the landscape. They also need an all-weather drive from public roads to their home or other areas of their property. If, however, driveway distances are very long, paving with concrete or asphalt may be cost prohibitive, and using alternative construction methods may be necessary to create an adequate driveway. It is important to build driveways on soil that can carry the weight of the traffic using it and, in northern and mountainous areas, on areas where snow does not accumulate. The sections that follow outline some of the considerations involved.

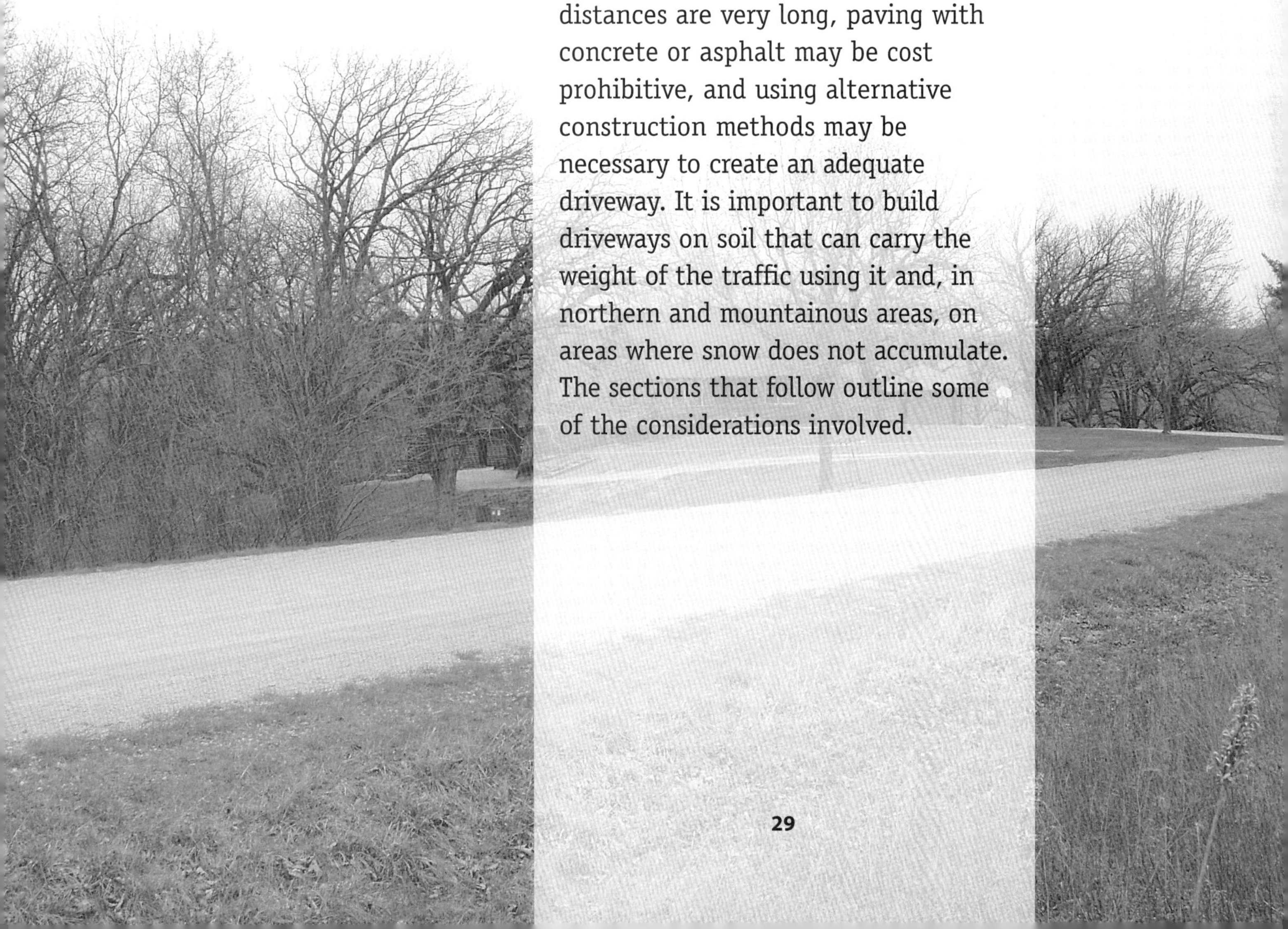

The Entrance Drive

Locate the driveway so it has an extended view in both directions at the point where it joins the public road. Hills, embankments, trees, glare from a low rising/setting sun position and other elements can obstruct the view of oncoming traffic. Locate drives either at the top of a hill or far enough from the top so drivers leaving the driveway have adequate visibility when pulling out into traffic (Figure 4-1).

The recommended distance to ensure safe visibility for vehicles entering and exiting a road with 60-mph traffic is about 1,000 feet (Table 4-1). Vehicles moving at 60 mph cover 88 feet every second. Fast moving traffic needs time and distance to avoid a vehicle exiting or entering the driveway. During the planning process, a practical way to evaluate the visibility of oncoming cars is to drive on the road at normal speeds and try to determine safe visibility, stopping, and acceleration distances. As a test, park a vehicle at the proposed driveway site and then drive the public road, noting the location from which the parked vehicle is first seen. The *decision sight distance* is the distance required for a driver to detect an unexpected vehicle or other hazard in the roadway, recognize its threat potential, and initiate action to stop or avoid the hazard.

Check with your local or state highway authority for information about and permits for driveway access to a public road. The driveway surface should be at least 16 feet wide. If large equipment or stock trailers will be using the drives, driveways should be wider (Figure 4-2). Vehicles waiting to enter the highway should be on a surface that is level or has a modest slope.

Provide at least 60 feet of driveway from the living area to an intermittently traveled public road. (The preferred distance is 100 feet.) Provide even more distance, and use

Table 4-1. Recognition and braking distances on rural roads.
Based on AASHTO *A Policy on Geometric Design of Highways and Streets, (The Green Book)*, 2001.

Travel Speed (mph)	Decision sight distance (feet)
45	675
50	750
55	865
60	990
65	1,050
70	1,105

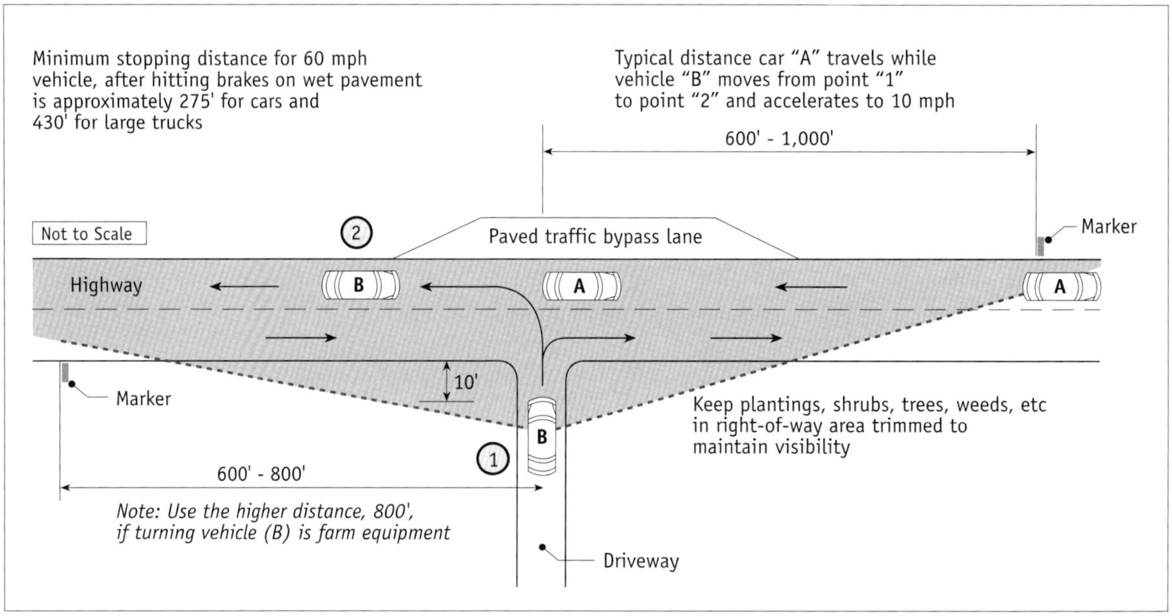

Figure 4-1. Traffic entry distance

trees to reduce noise from busy highways. While a greater distance is advantageous, balance that against the space needed for the driveway and the increased costs for construction and year-round maintenance. Branch drives off the main driveway can provide access to visitor parking and the family garage.

Driveways and Parking

Consider the following when planning drives and parking:

- Drainage to maintain the surface and control runoff.
- Area lighting for safety, security, and convenience.
- Space for maneuvering and parking.
- Surfacing for year-round access.
- Space for snow storage.
- Safety and convenience.

Plan curves with an 18-foot inside and 30-foot outside radius (Figure 4-3). Snow drifting, drainage, road right-of-way, utility easements, setback distances, ditch mowing, and surface maintenance are other concerns in driveway design and location.

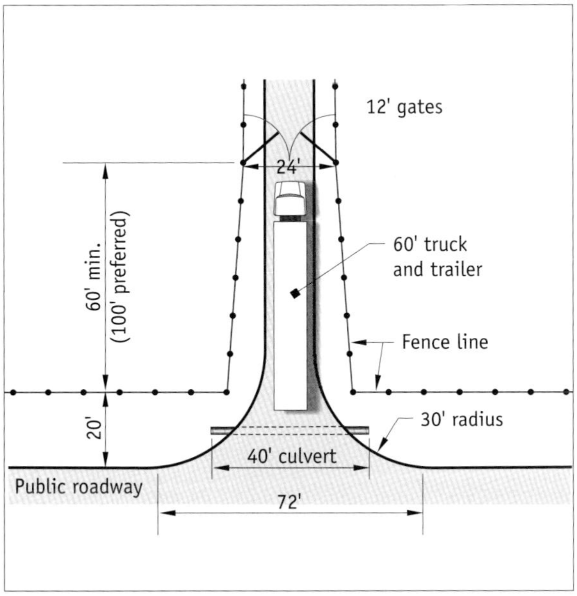

Figure 4-2. Driveway access to main road for stock trailer.

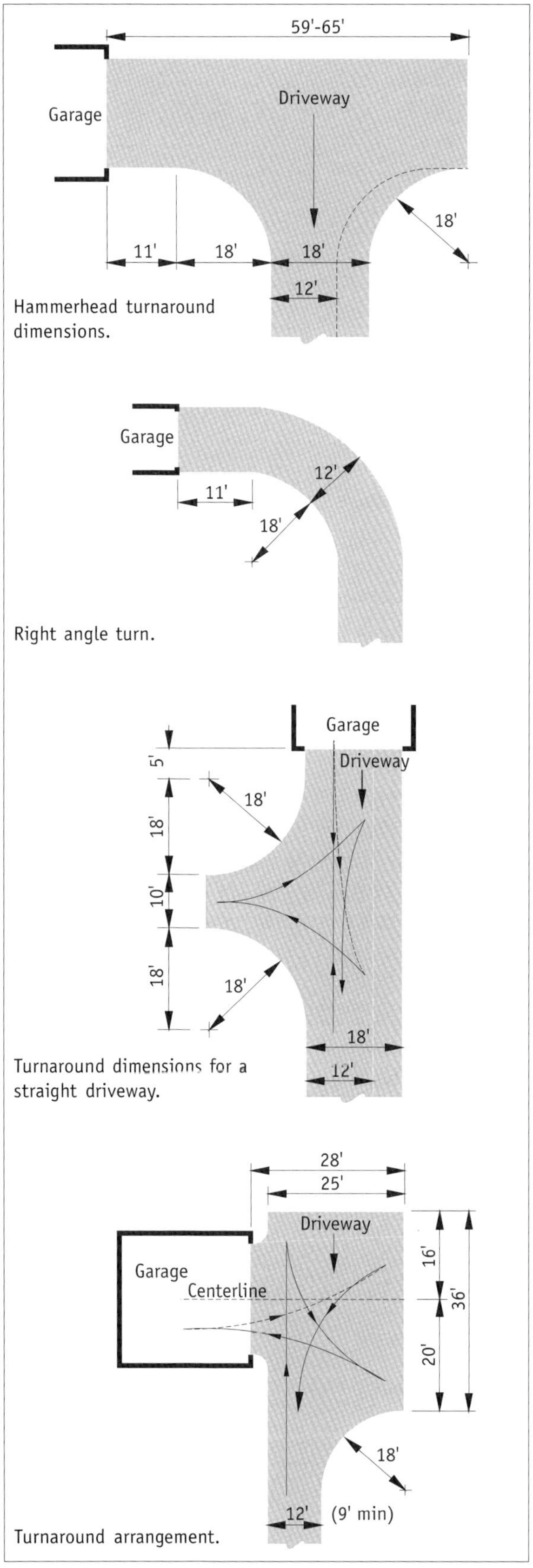

Figure 4-3. Turns, driveways, and turnarounds.

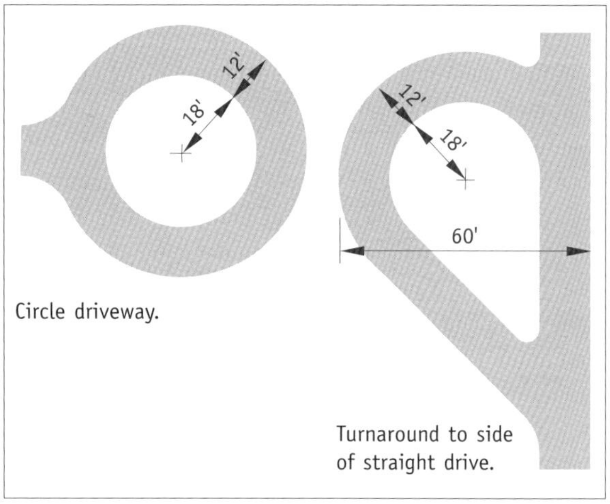

Figure 4-4. Driveway turns for cars and pickups.

All-weather Surfaces

Your selection of a driveway surface will affect its appearance, the construction expertise and equipment you will need, and the level of investment required. An all-weather gravel driveway surface is practical. A concrete or asphalt surface minimizes surface maintenance and contributes to a neat appearance. Concrete is the most durable material for heavy traffic areas and frequently scraped surfaces. Asphalt requires sealing every few years to prevent deterioration. Correct installation over a deep (at least 8-inch) compacted gravel base is essential for any all-weather driveway and heavy traffic drive surface.

Properly designed and maintained drainage ditches on each side of the roadway keep the base under the road well drained and maintain its load-bearing strength Figure 4-5. The subgrade (natural soil under the road, Figure 4-6) is important when developing an all-weather road. Most soils lose their bearing strength as moisture content increases.

Surface water can be kept below the subgrade by constructing ditches on either side of the road. The bottom of the ditch should be at least 1.5 feet below the surface of the subgrade. The cross section of the road ditch should be large enough to handle the expected flow of water without eroding the ditch bottom. The slope of the ditch and the ditch lining material should be designed to avoid erosion as well. Banks of the road ditch should be no steeper than 4:1 (4 inches horizontal to 1 inch vertical) to minimize the possibility of mowing equipment rollovers.

If the water table is likely to move into the subgrade, drainage tile with an unrestricted outlet is needed to keep the subgrade as dry as possible. Upslope drainage water should not flow over the road surface. Where flowing water must cross a road or driveway, use culverts to carry the water under the road.

Crown a gravel drive surface in the center of the road with a surface slope of half an

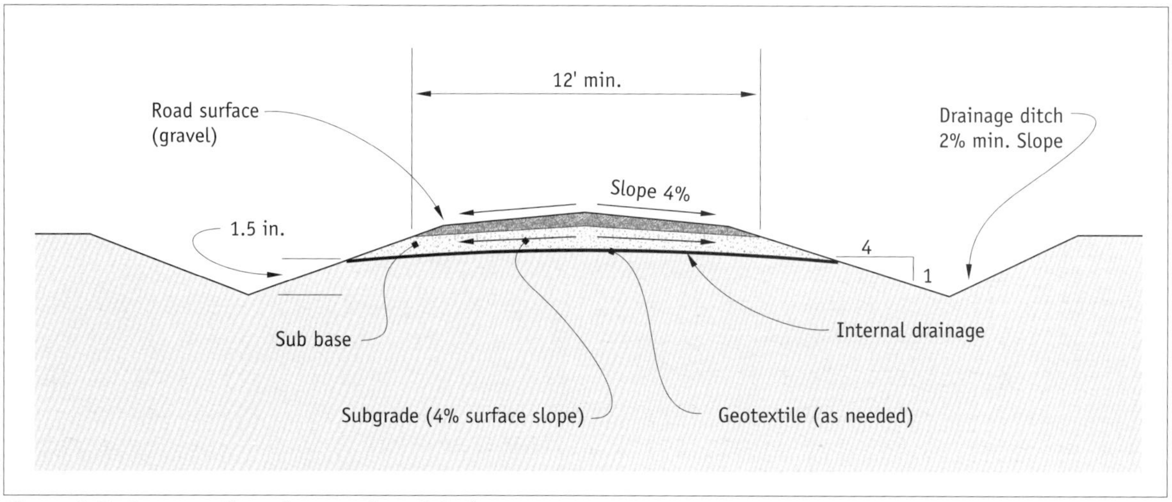

Figure 4-5. Cross section of a gravel road drainage system.

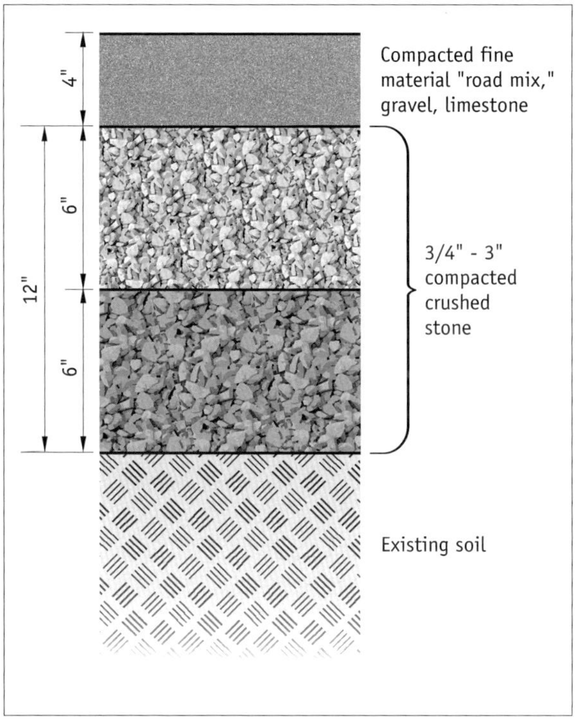

Figure 4-6. Cross section of an adequately constructed rural road surface.

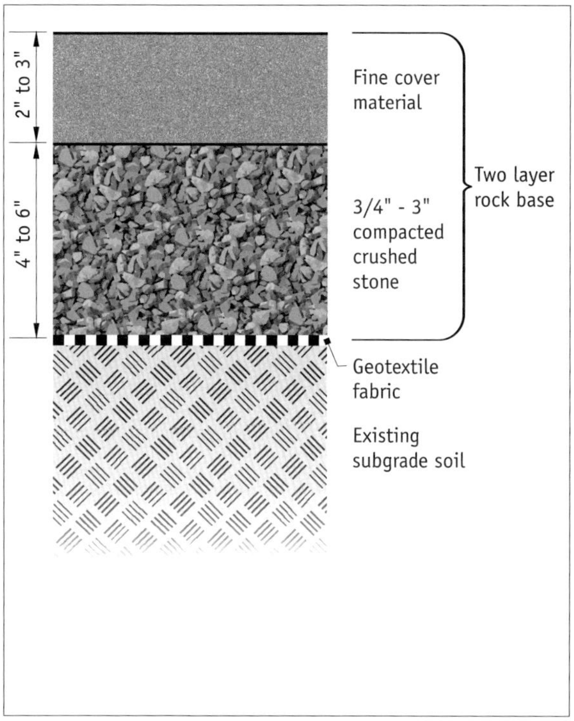

Figure 4-7. Cross section of adequately constructed rural road surface using geotextile.

inch per foot (4%) to encourage good surface drainage. Maintain the surface to prevent ruts, potholes, and gravel ridges at roadsides that can trap water and encourage infiltration into the subgrade. The road surfacing material should contain some fine soil particles (binder) to encourage runoff rather than infiltration. Wisconsin Transportation Bulletin No. 5 states that a good road surfacing gravel has the following characteristics (percentages based on weight):

- 40 to 80% hard stone, uniformly graded from 3/4 to 3 inches in diameter.
- 20 to 60% sand smaller than 1/4 inch.
- 8 to 15% fines (silt and clay).

Subbase materials can be well-graded gravel to breaker run (3/4 to 3 inches) with 0 to 10% fines. The thicker the subbase, the more uniform the load and lower the pressure exerted on the subgrade. Therefore, thicker subbase layers are needed for large, heavy loads such as loaded trucks and grain wagons. Subgrade soils containing a higher percentage of fines can increase water content in the subbase over time. Mixing soil with the subbase reduces its strength and limits its ability to internally drain water to the road ditch. Geotextile materials are often used as a barrier between the subgrade and subbase to preclude the mixing of subgrade soils into the subbase.

In recent years, geotextile fabric has been developed for all-weather traffic surfaces on farmsteads, Figure 4-6. Installed at a lower initial cost than concrete or asphalt pavement, geotextile fabric provides an economical pavement for livestock lots, less intensely traveled areas, and areas for stacking forage and piling manure. All-weather geotextile paved areas do require maintenance and can quickly wear from heavy traffic or scraping. They require similar compacted subbase preparation as asphalt or concrete pavement. A one- to two-layer rock base is spread over the carefully installed geotextile fabric. The fabric is porous so water passes through while gravel and rock that would otherwise be pushed down into the soil are held in place. Sub-surface drain tile may be installed to assist drainage and maintain load-bearing strength of the subgrade soil.

A fly-ash and Portland cement mix can also be used as a lower cost alternative to pave high-traffic farmstead areas. Fly-ash is a by-product of coal-fired, heat generation facilities and is available in different qualities. Higher grades with careful mix preparation, handling, and installation may replace a part of a concrete mix with little strength reduction. Mixing fly-ash with soil is sometimes used, but surface durability is a problem as well as corrosion problems from dry fly-ash if it blows onto metal roofing. Check with pollution control agencies or power suppliers to learn if unbound fly-ash is acceptable for use in your area.

The, *Farm and Home Concrete Handbook* (MWPS-35), provides details for concrete driveway construction, including subgrade preparation, selection, installation, reinforcing, and curing of concrete. *Using All-Weather Geotextile Lanes and Pads* (AED-45) explains geotextile fabric paving selection and installation and provides annual cost comparisons of concrete, asphalt, and graveled all-weather surfaces.

Visitor Parking

The layout and location of visitor parking should encourage its use (Figure 4-8). This involves providing walks, plantings, and signs to direct people where you want them to go. The layout should make guest parking areas apparent to visitors and encourage guests to park in the appropriate location. Paved walks, low shrubs, or a fence or hedge can help direct guests. Provide for drainage and lighting for evening visits, and have room for emergency vehicles and for parking. An inviting and safe walkway to the house is desirable.

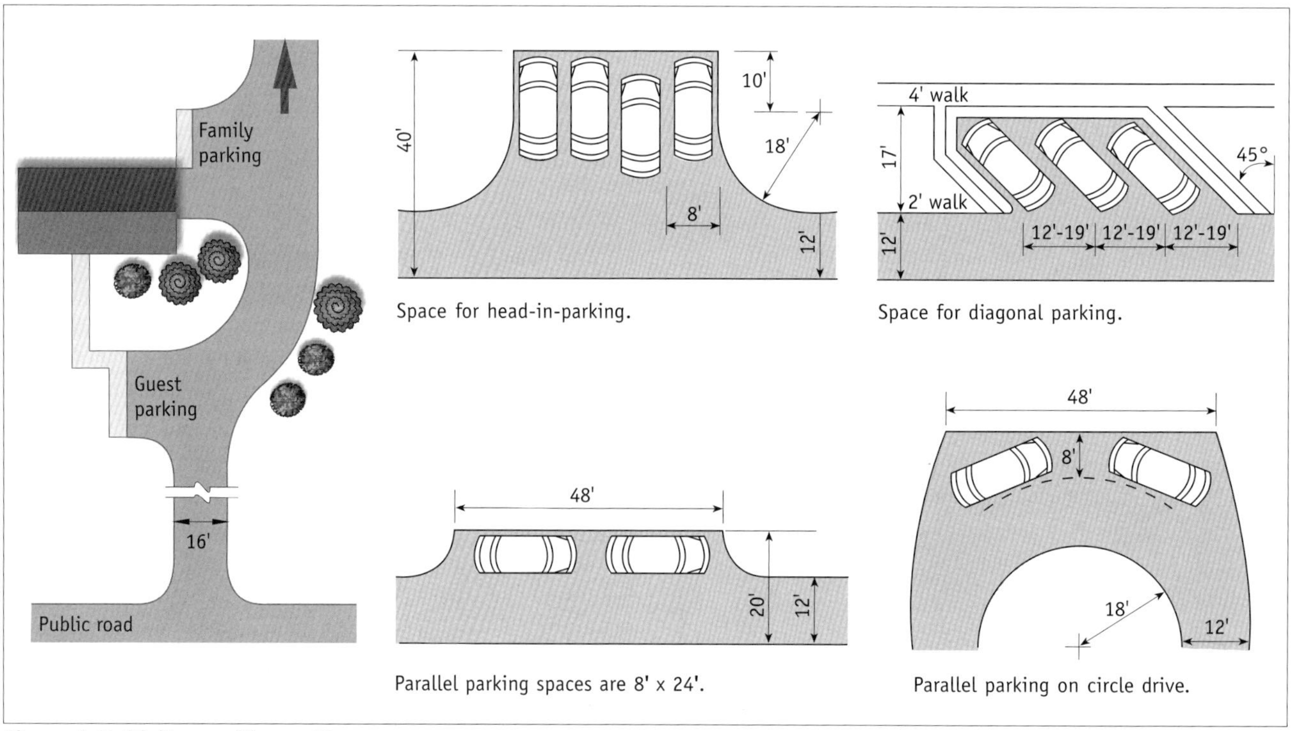

 Figure 4-8. Visitor parking options.

Other Considerations

In addition to the driveway and parking areas, other items to consider are culvert construction, sidewalks, and school bus stops.

Culverts

Culverts are used in areas where rainwater runoff needs to be transported under a driveway or farm service road. A minimum of a 12-inch diameter culvert should be used to minimize plugging. Many states, counties, and townships have permit requirements that specify the diameter, length, and installation process for a driveway culvert. The most common culvert is made from corrugated metal pipe 26 feet long with a 15-inch diameter. Note in Figure 4-2 that a 40-foot culvert is needed where the drive meets the main road. The construction and slope of the driveway as it leads into the road must also follow specific design guidelines. For example, the driveway must be level or slope away from the road where they join to prevent gravel or frozen runoff water from depositing on the road. Culvert aprons are required at each end of the culvert. This allows a gradual transition of soil slope from ditch bottom to the top of the culvert and minimizes the hazard to vehicles that accidentally leave the highway. Aprons also reduce soil erosion under the culvert.

Satisfactory culvert materials include PVC, high-density polyethylene sewer pipes, smooth-walled steel pipes with welded joints, and corrugated steel culverts. If you use sewer pipe, use smooth-walled pipe for culverts under farm roads. Culverts require periodic maintenance to remove sediment and debris that might clog them and lead to the roadway being overtopped. Use smooth-walled sewer pipe with joints capable of withstanding 40-psi internal pressure. Install culverts using the manufacturer's installation procedures.

Sidewalks

Residential sidewalks need to be designed to carry only a minimal load such as foot traffic and lawnmowers. From a safety and aesthetic standpoint, however, they should remain level or nearly so over time. Do not plant trees with shallow roots near a sidewalk, and place the sidewalk on well drained soil. The *Farm & Home Concrete Handbook* (MWPS-35) provides details for concrete sidewalk construction including, subgrade preparation, selection, reinforcement, installation, and curing of concrete. Construct sidewalks to the standard width of at least 4 feet; for many homes, a 5-foot width gives a more aesthetic balance when viewed from the road. Either width provides good wheelchair access.

School Bus Stops

Considerations in locating a school bus stopping location should include issues such as:

- Crash history in the vicinity of the school bus stop.
- Traffic visibility to and from the school bus stop.
- Provisions for passenger waiting facilities.
- Provisions for pedestrian facilities if students must cross the road before pick-up or after drop-off.

The optimum location for a school bus stop is on a straight section of the road and where there is a uniform grade with an unrestricted view of the approaching bus. Avoid placing a bus stop near sharp curves or just over hill crests. School bus stops should not be obstructed by vegetation or other objects. Visibility of the bus stop must be adequate from both directions of traffic.

Bus shelters provide children with weather protection (Figure 4-9). The shelter should not be placed where it restricts the sight distances of drivers leaving side roads or property accesses. It should be placed where the bus driver can see the waiting children in time to stop. The shelter can be enclosed on two, three, or four sides, depending on climate. Windows or other openings should be provided if needed for students to see the bus approaching. The shelter should be located where parents can monitor activity and where the bus driver has a good view of the shelter. Figure 4-10 shows plans for a small shelter.

Figure 4-9. Rural school bus shelter

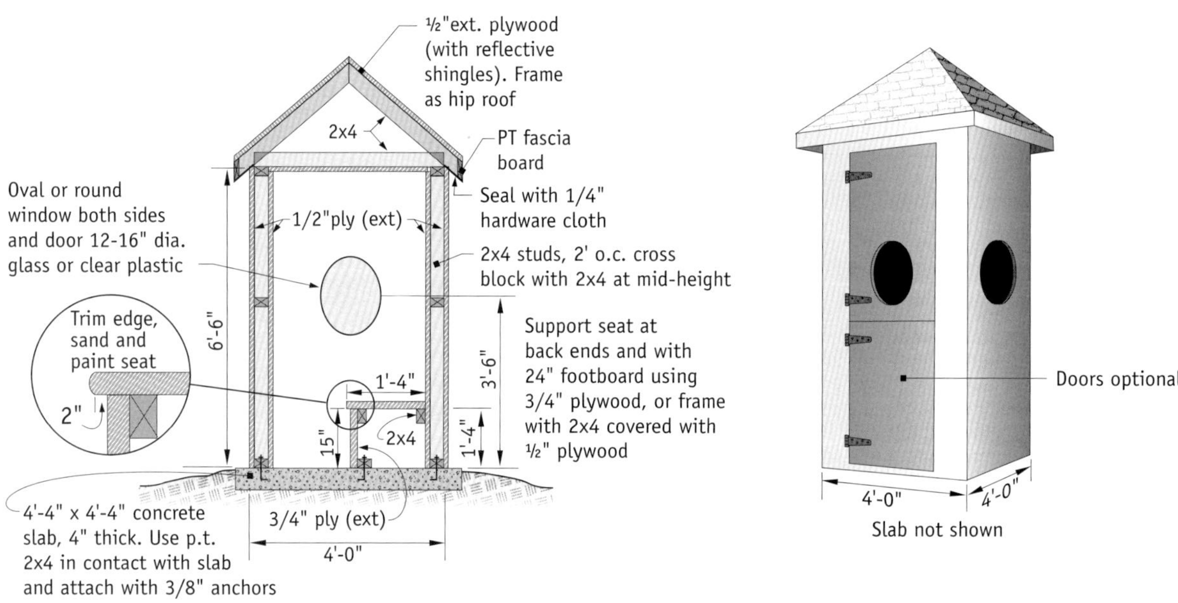

Figure 4-10. Construction drawing for 4-foot by 4-foot school bus shelter.

Resources

MidWest Plan Service, http://www.mwps.org

- *Farmstead Planning*, MWPS-2
- *Farm and Home Concrete Handbook*, MWPS-35
- *Using All-Weather Geotextile Lanes and Pads*, MWPS AED-45

The Ohio State University, http://ohioline.osu.edu/aex-fact0304.html

- *Using Geotextile Fabric in Livestock Operations*, AEX-304-97

5

Home Accessibility

A major, but sometimes neglected, aspect of planning a home is providing accessibility for people with special needs. While accessibility planning is usually thought of for handicapped, the same principles will make your home more accessible for everyone from young children to the aged. Consider curbs, door thresholds, aisle way clearances, and acceptable reach distances when planning for accessibility.

Design Standards

Universal design is intended to benefit everyone, not just the elderly or physically challenged. Even if you do not need some of these features today, you may appreciate having them later. It is usually less expensive to plan for future needs than to remodel later. If done initially, home accessibility features can often be built-in to the home design rather than appearing to be an obvious add-on if done later. The two major accessibility design strategies to consider are to choose a single-level house, and slope the entrance to your house rather than use steps. Steps may be hard for some of your friends and family to handle.

A paved surface is better for individuals who need a wheelchair, crutches or electric cart to maneuver. In some climates, a raised, covered walkway may be needed to minimize ice, snow, and drainage problems, particularly if buildings are used frequently. Access to buildings may require a ramp, bridge, or some kind of lift. A 1:12 slope is the recommended maximum slope for a fixed ramp (1:8 slope is permissible for existing buildings, ANSI A117.1-1998.) A wheelchair requires a minimum turning radius of 5 feet. A 32-inch.wide (minimum) clearance is needed for hand-operated wheelchair passage through interior and exterior doorways. Figure 5-1 shows examples of the space needed for wheelchair clearances.

A motorized electric opener may be needed to operate doors. There must be ways to enter and leave during an electric power outage or other emergency. A lever-action sliding-bar or self-locking gate is simpler to operate than hook, slide-bolt, or chain-type gate latches. It may be possible to utilize a solar or electric powered gate or modified garage door opener for gates or large doors.

There are a number of other design standards to be considered for accessibility, both inside and around your home. These include designs for ramps, hand rails, doors, and bathrooms. See Table 5-1 for details.

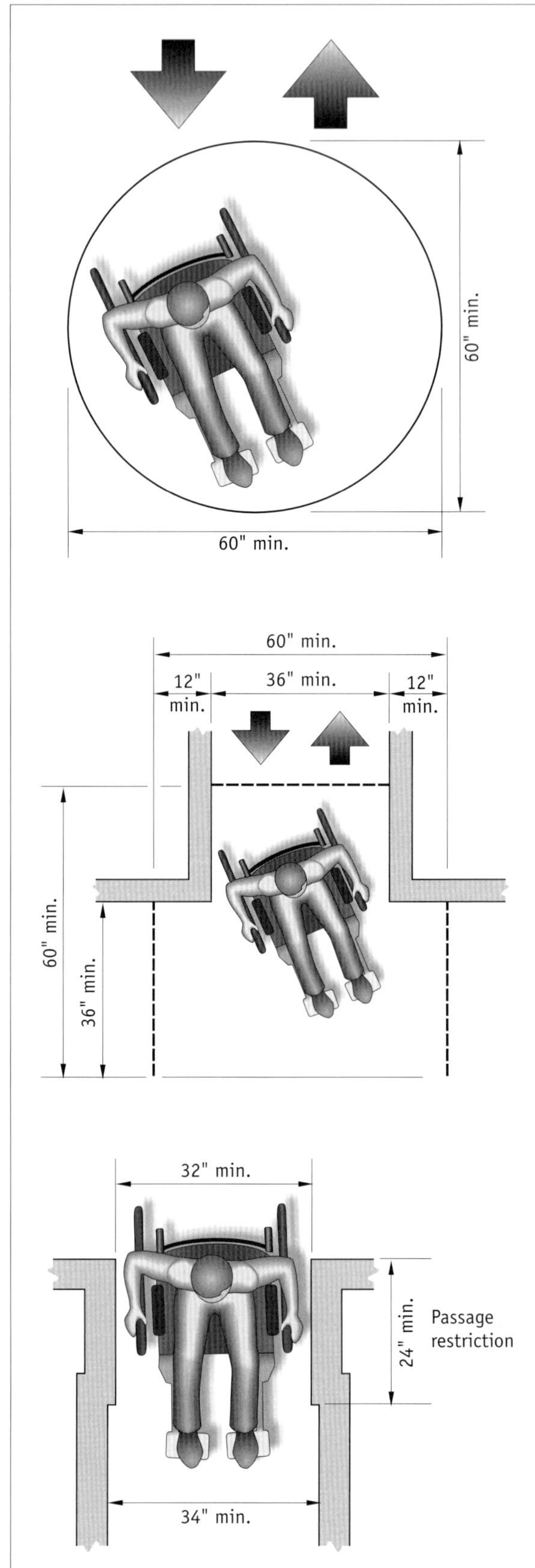

Figure 5-1. Selected wheelchair clearances
ANSI A117.1-1998

Table 5-1. Summary of Clearances needed for Farmstead Accessibility.
Source: *Breaking New Ground,* Purdue University
http://pasture.ecn.purdue.edu/~agenhtml/ABE/Extension/BNG/Outreach%20Program/outreach.html

Americans with Disabilities Act (ADA) Access Checklist	
Exterior	
Parking Space	13' (16') wide minimum (8' wide parking space plus 5' access aisle for cars, 8' for vans)
Curb Ramps	3' wide minimum, maximum 12:1 slope
Route use	Connect all accessible areas, 3' wide minimum, 5' x 5' passing area every 200' minimum, ramp if greater than 20:1 slope
Ramps	3' wide minimum, maximum 12:1 slope, 10:1 typical slope
Handrails	1 1/4" to 1 1/2" diameter, 34" to 38" to top of rail, 1½" from wall, required on ramps
Exterior Doors	32" wide minimum clear opening, bevel threshold if 1/4" to 3/4" high, lever-type latch, recommended maximum pull of 12 pounds.
Vestibule:	4' minimum between open doors, 4' plus door widths if doors open in
Interior	
Interior Doors	32" minimum clear opening, bevel threshold if 1/4" to 1/2" high, lever-type latch, recommended maximum pull of 5 pounds
Office	27" high, 30" wide and 19" deep knee clearance for desks/tables, 28" to 34" height for tables and counters
Water Closet	17" - 19" to top of seat, 1 1/4" to 1 1/2" diameter 36" long grab bar behind or 42" if on sides (extend side bar at least 54" from the back wall) if not in stall, paper dispenser 3' maximum from back wall and 19" off floor.
Lavatory	34" high maximum, lever handles on faucets, 40" to bottom of mirror
Toilet Stalls	60" wide x 59" long (56" if wall mounted toilet)
Drinking Fountain	36" high maximum spout outlet, 27" high knee clearance, 30" x 48" clear floor space
Telephones	30" x 48" clear floor space around phone, 48" maximum height for operable parts (front access) or 54" for side access
Protrusions	Includes phones, fire extinguishers, fountains, and other objects maximum intrusions of 4" into passage if leading edge of protrusion is between 27" and 80" above floor (for sight-impaired)
Access to Controls	15" to 48" off floor forward reach if forward access, 9" to 54" off floor side reach if parallel access

Figure 5-2 shows walkway and entry way dimensions for entry to the home that meet access requirements for wheelchair access. The Resources section lists both public and commercial websites where you can find more detailed information.

Figure 5-2. Walkway and house step dimensions.
Note: Railings have been removed to more clearly illustrate dimensions. A railing needs to be located on each side of the walkway and steps.

Resources

HGTV, Home and Garden TV,
- http://www.hgtv.com
- http://www.hgtv.com/hgtvdc_universal_design/ (note that the "spaces" are actually underscores)

Iowa State University, Extension,
http://www.extension.iastate.edu/universaldesign
- Universal Design & Home Accessibility

Advice and information for building an accessible home with information for both building and remodeling. The website includes information about entrances, bathrooms, kitchens, planning, design, and products that improve home accessibility.

MidWest Plan Service (MWPS),
http://www.mwps.org
- *The House Handbook*, MWPS-16

6

Outdoor Living Areas

Many people move to the country to enjoy more of the outdoors and are particularly interested in developing outdoor spaces to suit their needs. You may want to add a storage shed or extend your indoor living space by adding a deck or patio. You may also want to increase the wildlife around your home by selective planting and developing of a wildlife habitat. In any case, be sure to consider your options and costs when you are planning the layout of your outdoor living areas.

A great deal of variety can be created in an outdoor living area for a rural home. Although similar in some ways to those in urban areas, rural outdoor areas usually differ in extent. Living in the country, you probably will have more space to manage than if you lived in an urban area. You may be able to garden more and to create wildlife habitat. This chapter introduces you to general considerations for developing your outdoor living areas.

As you plan your outdoor spaces, consider how much time you have for upkeep. While there is less weekly maintenance on structures such as decks and patios than on a lawn, a decks and patios have increased initial costs and do require periodic maintenance. The type of outdoor living space you ultimately choose depends on the answers to several questions:

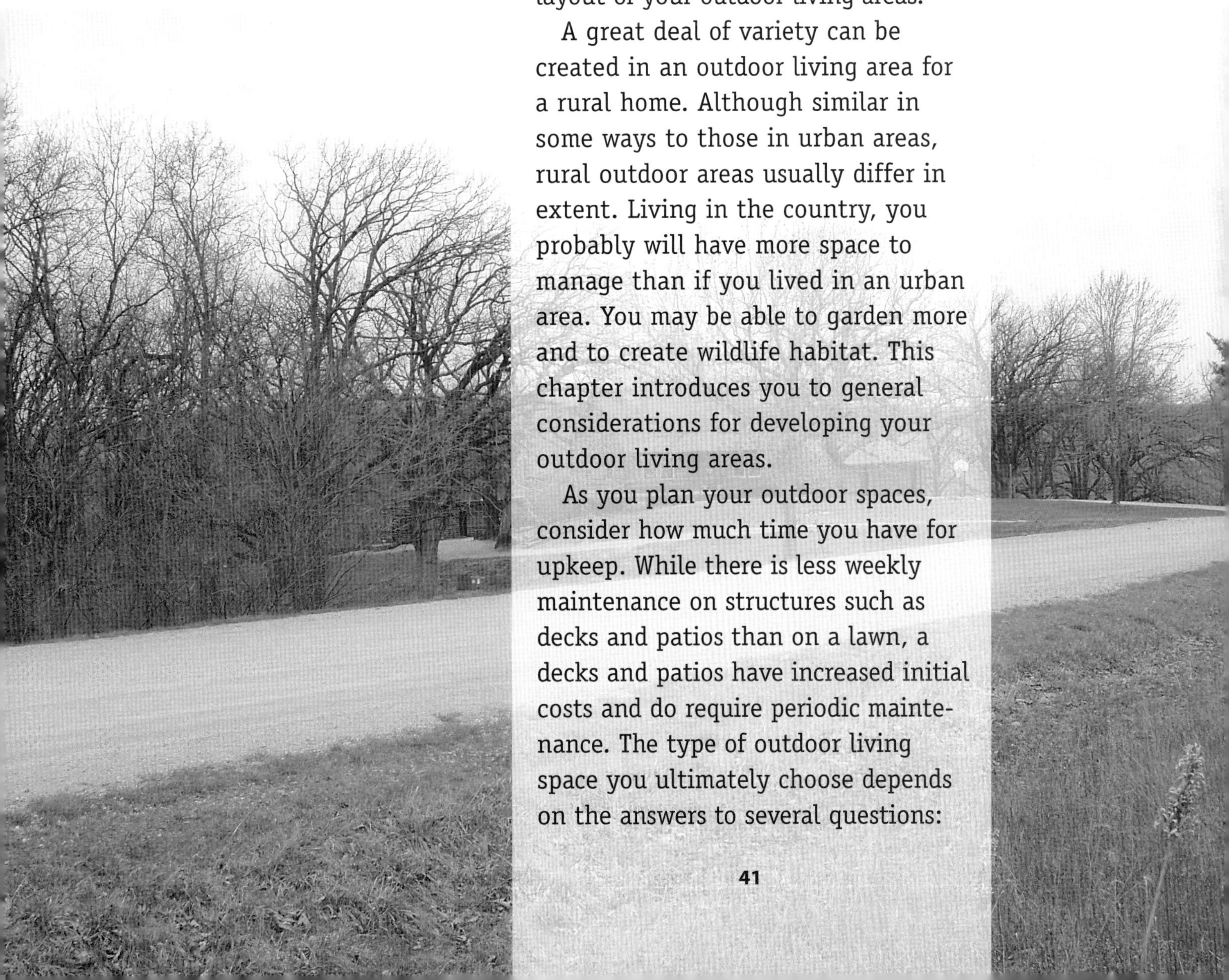

- How will you use your outdoor living area?
- Will you entertain frequently?
- Are there structures or features you want to hide from public view?
- Are there particular features to which you want to draw attention?
- Do you need easy access to the space from your home?
- Do you prefer to keep your outdoor space very private?
- Do you want to encourage birds and other wildlife by creating a wildlife habitat near your home?
- How much time and money are you willing to devote to the upkeep of your outdoor living space?
- How will outdoor improvements affect the resale value of your home?

There is a wealth of information about outdoor living available from your county extension service and from sources such as educational television and radio programming, the Internet, and from how to books found in your local libraries, bookstores, and lawn and garden stores. Take time to learn all you can, and plan before jumping into any significant project.

Lawns

How much lawn you have will depend on many factors, including:

- Your personal preferences.
- The placement and number of driveways and parking areas you have.
- The number and type of animals you have.
- The placement of your house and outbuildings.
- The location and size of your septic system soil absorption field.
- Local conditions, including the amount of rainfall, soil type, and temperatures.

If you have an extensive lawn, a riding mower is a necessity. Fertilizers can help grow a lush lawn, but using them requires you to mow more often. Chemical control of weeds must be done properly so as not to harm surface or groundwater quality or cause harm to domestic animals and to wildlife. Often, non-chemical options for controlling weeds and pests are available. Figure 6-1 shows a well-cared for lawn and outdoor living area.

Most rural residents simply mow their lawns as needed and do not spend the time and money trying to have a lush, green lawn as you might see in the city. When the weather gets hot in the summer, it is common to let the lawn dry, turn brown, and go dormant. Your lawn will green up again when adequate rain returns. When planning the size and design of the lawn and garden, realistically estimate the time each week that you are willing to spend on maintenance. You many choose to turn a portion of your space into a woodlot or wetland instead of a lawn. Likewise, shrubs, tree leaves and limbs, and flower and garden areas may take a significant amount of time to maintain, especially on a large property. In general, rural homeowners restrict the lawn area to a maximum of about an acre to minimize maintenance concerns.

You may wish to use native plants for landscaping since they are well suited to the soils and local weather conditions. The EPA has a resource, *"Landscaping with Native Plants,"* at http://www.epa.gov/greenacres/nativeplants/factsht.html.

Figure 6-1. An attractive lawn provides outdoor living spaces and improves the appearance of your home.

Decks and Patios

Decks and patios, particularly those with barbeques and/or fire pits, can increase your use and enjoyment of your yard. Wooden decks are popular because materials are easily available, and they are relatively easy to install and maintain. Patios are built at ground level and constructed of concrete, stone, tiles, or other weather-resistant materials. The deck or patio should be easily accessible from the house. In rural areas, open fire pits are often built into patios. They can be fun to watch, will take the chill off a cool evening, and extend the length of the season in which you can use your outdoor space. Use a commercial unit or construct with fire safety in mind.

Garden Areas

When developing flower and vegetable gardens, provide adequate fencing to exclude children and wildlife. Gardens provide a pleasant means of exercise, beauty, and aesthetic pleasure. Gardens also provide fresh, nutritious food. The best location for your garden depends on many considerations, including soil condition, available sunlight, garden size, availability of water for irrigation, and accessibility (distance to your house). You may also choose to plant a close-by "kitchen garden" for frequently used herbs and tomatoes. A number of good resources are available for those who have not gardened before. Master Gardeners teach practical gardening classes in many areas around the U.S. and are an excellent source of local information. Contact your county Extension Educator to locate a Master Gardener near you.

Flowers, nuts and berries, and brilliant foliage are pleasing in landscape plantings and attract birds and butterflies. A good mix of plantings can provide wildlife habitat and enjoyment year-round. Shrubs can be used for foundation plantings, screening, and division of different areas. Consider your soil, moisture conditions, and sun exposure as you choose what to plant.

Pruning is an important part of garden maintenance. Do not, except in the cases of clipped hedges, rely on pruning to change the natural shape or size of a plant. Recommended pruning times for shrubs varies according to the time of blooming. Early flowering types, like the forsythias, lilacs, and quince, form their flower buds on last year's wood and should be pruned after flowering. Later flowering types, such as mock orange, butterfly bush, and the hydrangeas, can be pruned in the spring.

Consult with neighbors to learn of the wildlife pests in the neighborhood and which garden plants and shrubs they like to browse. Either don't plant those species, or devise a plan for protecting them from the pests.

Trees and Shrubs

Trees and shrubs can beautify your property, shade your home, and reduce the force of the wind in winter. On a hot summer day, the temperature under a large shade tree can be 10 degrees lower than the temperature in the sun. If you have the space, it is best to plant trees far enough away from the house so that branches do not hang over the roof. Overhanging branches can lead to mold buildup on the roof and siding and to leaves clogging gutters and downspouts. Overhanging branches also can provide a convenient entry to your attic for rodents and other pests.

Strategically planted trees and shrubs provide protection against the wind and can diminish both its direct and indirect effects. Windbreaks reduce the force of the wind on the exterior surfaces of buildings and thus the amount of cold air that enters the home. They also can be placed to reduce snow drifts near your home. Protecting working areas around the home with windbreaks makes tasks such as cutting firewood, working on equipment, or feeding livestock safer and more comfortable. Figure 6-2 shows a typical configuration of a shelterbelt. A farmstead windbreak can also improve living conditions by screening undesirable sights, sounds,

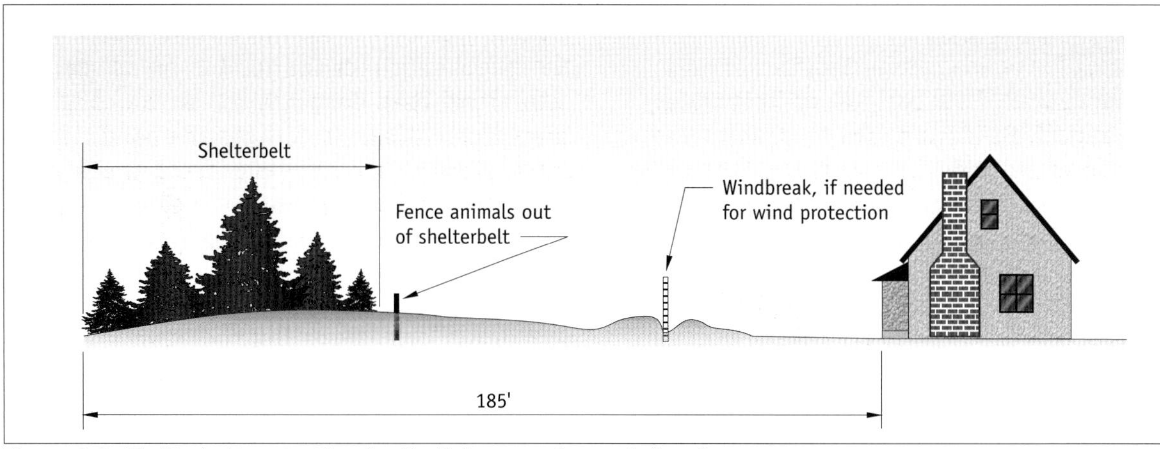

Figure 6-2. Shelterbelt protecting family living area from wind and snow.
Snowdrift depth and length depend on windbreak type, height, and width; wind velocity; and amount of snow. Generally most snow accumulates within 185' of the windward side of the shelterbelt.

smells, and dust. Windbreaks reduce traffic noise from nearby roads and highways by deflecting sound off of large branches and tree trunks and by absorption of sound by leaves, needles, twigs, and smaller branches. In addition, trees and shrubs can mask undesirable noises with the more desirable sounds of the rustling of leaves or the singing of birds attracted to the windbreak.

Recreation is part of a healthy family life. Activities such as picnics or barbecues, playing ball with children, or working in the garden are far more enjoyable behind the shelter of trees on a windy day.

Before planting you must decide upon the best location of the windbreak. Consider the location of your buildings, the direction of prevailing winds, and your soil type. For the best windbreak, choose trees and shrubs suitable for your soil and climate conditions. Species native to your area will adapt most easily and are usually the best choice. With a mixture of deciduous trees, conifers, and shrubs, the windbreak will provide protection all year.

Consider what you want your outdoor living area to look like, and add species that have flowering, fruiting, form, color, or texture characteristics that beautify your home. Enhanced aesthetic quality and a more natural look can be achieved by using curved rows or adding clumps of trees or shrubs to the planting. See Chapter 8 for more information on windbreaks.

Plantings of trees and shrubs that block the view of your house from public roads can increase your privacy. However, extensive plantings can also reduce the ability of your neighbors (and the county sheriff) to keep an eye on your property. If you choose to plant trees and shrubs that block the view from the road, consider leaving a couple of open views of your house from the road. See Chapter 8 for additional information about security of homes in rural areas.

Ponds

A well placed pond can add value to the farm, be a source of family recreation, attraction for wildlife and even serve as a source of water for livestock, the home, and fire protection (Figure 6-3). Several steps are involved in constructing a successful pond. The first is to make sure the site is suitable in terms of soil and geology so the pond can hold water. Next, verify that the source of water is adequate and that the pond can be properly managed. Visit the local NRCS office if you are considering constructing a pond. State extension services and state wildlife and fisheries authorities can provide management information for ponds.

Pond Construction

Water for ponds can be provided from a stream or well, from springs, or from watershed runoff. Ponds supplied by runoff from upslope watersheds are constructed either by excavation or by construction of berms (earthen dams) around one or more sides of the pond to trap the runoff or by some combination of the two methods. In order to ensure that soil in the area will hold water, soil borings and a geological examination of the area are needed to ensure that areas underlain with limestone, gravel, shale, or similar materials are avoided. If the soil is not adequate, adding a bentonite clay sealant or a plastic pond liner may be necessary. Swampy areas are poor choices since they are hard to drain and provide a weak foundation to support the weight of any dam. If water from a flowing stream is used to provide water to the pond, construct the pond in an adjacent area rather than damming the stream, and then divert stream water into the pond. Most states have legal restrictions about removing water from a flowing stream, so check with your state natural resources agency.

Figure 6-3. Farm pond used for recreation and a water source.
USDA photo gallery.

Save topsoil from any excavation and place it on the sides and dam slopes to support vegetation and reduce erosion. Pond dams should be designed and constructed by qualified professionals. Some states require a dam construction permit if the volume of water held behind the dam exceeds a certain amount. For example, North Dakota requires a construction permit if the dam retains more than twelve and a half acre-feet of water. Pipes extending through the dam should be installed very carefully and preferably only at the time the dam is constructed. Your local NRCS should be able to assist or direct you to local professionals.

When constructing a pond fed by a watershed, a good rule of thumb is to make sure that water runoff enters and leaves your property at the same points it did before the pond was constructed. This should prevent any water access or drainage disputes with neighboring landowners.

Protecting a Pond's Watershed

Most ponds collect runoff from an adjacent watershed. If possible, locate a watershed pond so that the watershed is under your ownership or control. The use that is made of the watershed determines the quality of water in the pond. If manure or fertilizer or other materials are applied to the watershed at a rate that is higher than can be used by vegetation growing there, or applied in a manner so that it runs off before it can be utilized, some of the applied material can end up in the pond.

If runoff from the watershed is contaminated, it may be possible to divert the portion of the watershed that is adding the contaminants. If this is not feasible, place a buffer strip or wetland below the watershed to clean the runoff before it enters the pond. Minimal and careful use of fertilizers, pesticides, and fungicides can help protect the water quality around your home.

Watershed size needed to sustain a pond depends on soil type, average rainfall levels, and vegetation grown in the watershed. Consult your local NRCS for design requirements in your area.

Pond Safety

If the pond is used for swimming, rules should be established, posted and followed, just as for any body of water where people swim. Make sure that swimmers have received proper training and do not swim alone. Be sure that children have adequate adult supervision. Remove any submerged obstacles that could injure or entangle a swimmer. Construct a dock and ladder to make entering and exiting deep water in the pond simpler. Provide a rescue post that is clearly visible from all points around the pond, and place a 12- to 14-foot rescue pole, nylon rope, and lifebuoy at that point. The rope should be long enough to reach across the pond, with one end attached to the buoy and the other to the post.

Using a Pond for Fire Control

A farm pond can be an excellent source of water for firefighting. Although many fire departments are equipped to draw water directly from the surface of a pond, a better way is to construct a dry hydrant in the pond embankment if the terrain allows. See Chapter 8 for details about this type of installation and for additional information about using a farm pond for fire control.

Animal Damage to Ponds and Surrounding Areas

Most animal damage to ponds can be prevented or controlled relatively easily. Large animals should be fenced out of ponds (with the fence at least 60 to 100 feet from the pond) if the pond could overflow into a waterway. Large animals can damage the pond embankment with their hooves, increasing erosion and silt sedimentation in the pond. Wading animals stir up mud in the pond and can disturb fish habitats. If the pond is to be used to supply water to the animals, plan for that before the pond is constructed and, if possible, provide water via a gravity feed to a stock watering tank located below the pond dam with a gravel or concrete pad around the waterer. At some sites, it may be necessary to pump water from the pond to the point of use in order to provide water to animals.

Muskrats are common in some parts of the country, primarily in the Midwest. Normally, muskrats construct dome-like homes of sticks and other debris in shallow water, where the top of the structure protrudes 6 to 12 inches above water level. The animals normally enter the nest below the water's surface with the nest located just above water level. When constructed in shallow water, muskrat homes seldom cause much concern; in fact, ducks and geese often use the tops of the structures as nesting areas.

Muskrats can cause serious problems if they begin to burrow into dams. These burrows are similar to the layout of the nests that muskrats construct in shallow water. Animals enter their burrows under the water level with the burrow sloping upward to the nest. Constructing the pond dam with at least two feet of freeboard and a dam least 12 feet wide with 6- to 12-inch large stone riprap at water level should minimize the possibility of muskrat damage or of a burrow reaching entirely through the dam. There have been instances however, where vehicles traveling over the dam breaks through into the nest. This can negate the protection of the freeboard and can create an overflow at that point, resulting in erosion and failure of the dam. Muskrats can be easily removed by trapping, and muskrat pelts have value. Check with your local wildlife regulatory agency to determine any restrictions on trapping and how best to remove wildlife pests from your area.

Other pests of concern around ponds include geese and ducks, which in large numbers are a nuisance and can result in high waste loading of the pond and can even affect plant and fish life in the pond. If geese are allowed to nest, they will be difficult to remove from the area. Again, contact your local wildlife agency for restrictions on hunting and animal removal in your area. Allowing grass to grow high around the pond, and even in a corridor between the pond and a feed source, such as a corn field, will increase bird and other wildlife in an area. Conversely, keeping the grass around the pond closely mown will reduce wildlife around the pond. Keep trees and shrubs away from the pond to prevent root damage to the pond's berms. Also, trees located near the edge of the pond drop leaves and other organic debris into the pond. Tall grasses can inhibit ducks and geese.

Other pond pests include beavers, which can damage nearby trees used to construct their homes in the pond and groundhogs, which can burrow into the dam, usually from the downslope side. In warmer, wetter areas of the U.S., water snakes and alligators can take up residence in and around ponds.

Mosquitoes can be a problem in small or very shallow ponds that have a significant floating pond cover or stagnant water that lacks a significant fish population. Reducing nutrients entering the pond should reduce vegetation growing in and around the pond and reduce mosquito larvae. Artificial aeration can help by reducing stagnant water. The use of chemicals to control mosquitoes is best left to a certified pesticide applicator.

Leaking Ponds

The amount of normal water loss from evaporation will depend entirely on location and local climate conditions, but any time a pond level drops more than expected (for example, more than 6 to 12 inches per month in the Eastern Corn Belt), suspect a leak. It is possible to attempt to stop the leakage by uniformly spraying slurry containing bentonite clay on the pond's surface under still water conditions and allow the clay to settle to the bottom of the pond. Bentonite clay particles swell to several times their size when wet and can form an effective seal. Success from applying clay to the pond surface is not always effective however.

A much more successful method is to first drain the pond and allow the soil on the bottom and sides of the pond to dry. Bentonite, as a dry power, can then be tilled into the top several inches of soil. A rate of 1 to 3 pounds per square foot of soil surface is typical, but the actual application rate needed depends on local soil type. The cost of bentonite is greatly dependent on hauling distance, so a local source of the clay can make the use of bentonite relatively inexpensive. Conversely, if it is not available locally, transportation costs can make bentonite use prohibitively expensive.

A third method of preventing a leaky pond is to use a plastic or rubber liner along the bottom and sides. These geomembranes are relatively expensive but can be highly effective. After the pond is drained, or before it is used, the membrane is installed and then covered with several inches of sand and soil to hold the membrane in place and protect it from punctures. Liners are normally placed side-by-side in several wide strips, with the strips fused together where they join, and the liner is anchored at the top of each side of the pond.

Wildlife Areas and Wildlife Preservation

Trees and shrub plantings encourage wildlife to visit your property, especially birds, rabbits, and squirrels. Windbreaks provide sites for nesting, feeding, and breeding for many birds and animals. Additionally, windbreaks furnish a safe, linear corridor for wildlife movement between feeding, watering, and resting sites. Wildlife management is often thought of in terms of protecting, enhancing, and nurturing wildlife populations by providing the habitat needed for their well-being.

Improving the quality of land increases wild bird populations. Bird habitats can be improved by managing food and cover plants. Wildlife habitat management is becoming even more important with the increasing loss of farmland and natural landscapes in many areas. The trend toward larger farms with fewer brushy fencerows between fields, fewer woodlots, and shrinking wetlands has also reduced habitats available to varied bird populations. Expansive, close-cropped lawns, street plantings with the same kind of trees, and vast fields of grain without shrubby fencerows all have the same negative effect on wild bird populations. Monotonous plant communities lead to similarly monotonous bird populations. As a rule, land development simplifies plant and animal communities, resulting in less diversity of plants, birds, and other animals.

Some bird species have vastly extended their ranges during the past 50 years because people have created and protected needed habitat. Bird species originally regarded as southern, now reside permanently as far north as southern Canada and have established themselves throughout much of New England and the Midwest. Given the proper habitat, you may notice birds that used to migrate south now staying through the winter. Their winter survival has been enhanced by feeders, habitat protection, and hunting restrictions. The U.S. Fish and Wildlife Service estimates that Americans spend a billion dollars each year on birdseed. Plants such as forsythia can act as an excellent habitat for many birds. Rhododendrons and other evergreen shrubs and trees provide good nesting sites.

The number of animals that can live within any one habitat area is determined by one of several restrictions known as limiting factors. These commonly include food, cover, water, and nesting sites, but other factors such as parasites, predators, display areas, or a lack of singing posts may also limit bird populations. Limiting factors can change from one season to the next. Food may limit numbers in the winter but not in the summer. Likewise, cover may be sufficient in the summer but not in the winter. Habitat improvement through vegetation manipulation is often slow, but for those with patience and an interest in landscaping and gardening, there will often be a long-term benefit of planning and planting wildlife habitat. It is important to create and protect plant communities that provide food, cover, and nesting sites for a variety of species of birds. Artificial nest boxes, platforms, and watering devices can also make a real difference when there is a lack of suitable natural nesting sites or water supplies.

Encouraging wildlife, however, can also bring undesirable pests to your home, so wildlife should be carefully managed to encourage the particular species you want while not creating a habitat that encourages unwanted species. Wildlife damage control is increasingly necessary because of expanding human populations and intensified land-use practices. Wildlife is generally considered a pest if it causes damage to your property, garden, or trees. If you feed cats or dogs outside you may have problems with raccoons that are attracted to the food. See Chapter 7 for suggestions about dealing with wildlife pests.

Resources: References

Do it Yourself network, Cable/satellite TV about home and garden projects,
http://www.diynetwork.com

Gardening Publications—Indiana,
http://www.ces.purdue.edu/new/gardenmailer.pdf

HGTV, Home and Garden TV,
http://www.hgtv.com

Landscaping with Native Plants,
http://www.epa.gov/glnpo/greenacres/wildones

Management of Ponds, Wetlands, and Other Water Reservoirs to Minimize Mosquitoes Water Quality, WQ-41-W.
Brent Ladd and Jane Frankenberger, Purdue University

Master Gardeners & Jr. Master Gardeners,
http://www.ahs.org/master_gardeners

National WHEP site, the national Wildlife Habitat Evaluation Program involving 26 state and over 10,000 4-H youth,
http://www.whep.org/Online

Ohio Pond Management Handbook,
A guide to managing ponds for fishing and attracting wildlife.
M Austin, H, Devine, L. Goedde, M. Greenlee, T. Hall, L. Johnson and P. Moser, Ohio Division of Wildlife

Online Gardening Publications—Indiana,
http://www.ces/purdue.edu/new/gardenmailer.pdf

Pond Construction: Some Practical Considerations.
Louis A. Helfrich and Garland B. Pardue Department of Fisheries and Wildlife Sciences, Virginia Tech
http://www.wildlifemanagement.info/publications/fish_ponds_33.htm

Popular Mechanics,
http://www.popularmechanics.com/

Prevention and Control of Wildlife Damage,
ISBN: 0-9613015-1-1 (book), 0-9613015-2-X (CD-ROM), 0-9613015-3-8 (book & CD-ROM),
University of Nebraska, first printing, 1994.

Rural Security Limited,
http://www.ruralsecurity.sageweb.co.uk

***Rural Security Planning**, PPP 64,*
http://www.btny.purdue.edu/Pubs/PPP/PPP-64.pdf

Sealing Leaking Ponds,
Martin W. Brunson, PhD
Adapted from *Ponds–Planning, Design, and Construction*, USDA Soil Conservation Service Agriculture Handbook Number 590
http://www.wildlifemanagement.info/publications/fish_ponds_14.htm

The Audubon Society. Books about Attracting and Feeding Birds,
http://www.audubon.org/market/licensed/backyardgardeningbooks.html

Resources: Construction Plans

The USDA plan service was active in developing a variety of farm plans until the 1980's. North Dakota State University was a repository of these plans and scanned them to make electronic versions of these plans available on their Agricultural and Biosystems Engineering Extension website, http://www.ageng.ndsu.nodak.edu/. Look under Extension, Building Plans. Plans are divided into the following categories: beef, construction, crops, dairy, horse, housing, machinery, poultry, sheep, swine, and a miscellaneous category. These plans were developed several years ago and should be checked by a professional engineer to ensure the plan meets building codes for your area. Plans dealing with the following, along with many others, can be found at this website.

TITLE
■ Barbecue pit, 3x12x3, conc. Block+firebrick, 8x13, c. Slab
■ Beehive & frame, 22" wide, plywood
■ Bike rack, 5 bikes hanging, welded sq. Tubing
■ Boat dock + shelter, 16x14, poleframe, gbr w/hoist & bench
■ Chore clothes cabinet, 1.5x2'x7', 1 drawr, shelvs & door, ml
■ Club house, 25x42, kitchen, studwall, gableroof, f, rwf
■ Carport, 20x34, shop & stg, 1/2-monitor roof
■ Dog house, 4'x4'x3', insulated, shr, skids
■ Dog houses, 2'x3', 4 types, gableroof, shedroof ('77)
■ Energy efficient landscape houseyard design

- Fruit & vegetable storage
- Fruit drier, 1.5x2x2, 1.5 kw, 60 cfm, portable
- Fruit drier, 2x2x2, 5 trays, 2 kw, 80 cfm, elec. Details
- Fruit/veg. Strg, 22x38, belowgrade, reinf c., end access, gv
- Garage layouts, approaches & cross-section, 1 & 2 cars
- Greenhouse & strg, 10x14x8, flatroof, movable
- Greenhouse benches, 4'x12', 5 types, wood stepbench
- Incinerator, 3'x4'x3' + chimney, brick const.
- Lawn tool storage cabinet, 4'x7'x1', wall mounted, plywood
- Low cost beehive, 1" & 2" lumber
- Nursery plant lath house, 23'x48'x5', rigid frame, pb found.
- One car garage
- Outdoor camp stoves/grills, reinf. Conc or stone or brick
- Outdoor grills, 3x4x3, 4 types, masonry & rock const.
- Outdoor porch deck, 14x16, post-beam supports 10' oc
- Outside fireplace grill, brick & conc, wood picnic table
- Patio bench, table & children play table
- Patio chairs, wood, 4 designs
- Picnic shelter & strg, 24x36x10, pole frm, truss rafter
- Picnic shelter w/kitchen, 24x36x10, truss rafter, stw, cf
- Picnic table, 3 1/2 x 6, attached seats, wood plank
- Picnic table, 3 1/2 x 6, movable seats, wood plank
- Plant growth chamber, 2x4x7, on legs, materials list
- Plant potter & stand, 1'x3'x1.5', sawhorse type
- Plant propagation unit, 3'x6', lights, watering system
- Playground equipment: ladder, wagon, swings, teeter, sandbox
- Playground layouts, 7 games incl: baseball, football, b.ball
- Purple martin birdhouse
- Rabbit barn, 10'x10', post-beam, elev. Cages
- Rabbit barn, 8x8x6, gableroof, studwall, rwf, skids
- Rabbit hutch, 3'x8' cage between posts, shr
- Roadside market, 30x60x7 w/refrig rm, tr, cb, cf, insul.
- Roadside stand, 4 layouts, 12x24, 16x24 and 16x36, gbr
- Screened & sheltered bbq pit, 13x25x9, poleframe gbr, cf
- Serving cabinet, 2'x2'x3', top drawer, shelfs&doors, movable
- Sheltered bbq pit, 13x19x9, poleframe, gableroof, cf
- Shop workbench, 2'x6'x3' and 2 sawhorses
- Slot ventilation ceiling/wall air inlets, 3 types, adjustab
- Solar beeswax extractor, 2x3x1, plywood
- Solar collector, clear poly, 20x96x4 for grain dry fan ('80)
- Solar fruit drier, 4'x4'x1', 4 trays, legs

- Spring-water collector/storage, 4'x4'x6', r. Conc., shr
- Stair & railings for outdoor porch deck
- Stool, 1.5'x2.5', 4 legs, wood
- Storage bldg, 8x8x8, shedroof, studwall, cf, materials list
- Storage building, 10'x12'x12', gambrel roof, cf
- Storage building, 12x12x8, shedroof, studwall, skids
- Storage cellar, 14x20x9, under gableroof, stw workshop, gv
- Storm or storage cellar, 8'x11'x7', reinf. Conc., steps
- Storm shelter, 8x12x8, reinf conc, flatroof, aboveground, cb
- Trash burners, 3x3x3, 3 types, masonry & concrete
- Utility &food strg cabinet, 15"x2'x7', 1 drawr, shlvs&dr, ml
- Utility cabinet, 1.5'x2'x7', 1 drawr, shelves & door, ml
- Vegetable cellar, 8x8x8, reinf conc, gravity vent
- Vegetable keeper for heated basement
- Vegetable storage rm, 10x10x8, stw vent & insul, door const
- Walk-in refrigerator, 8'x12'x8', shedroof, skids
- Wall cabinets for food preparation; framing, drawers, doors
- Wall storage cabinet, 2'x4'x7', doors & shelves, ml
- Windbreak fence, 14'x14'x14', spaced boards, skids
- Windbreak fence, spaced-boards 8' high (dickinson)
- Wood box w/kindling & paper drawers, 24"x27"x30", ml
- Wood post pres. Trtmnt tank, 1x1x18, hngd cover, metal lined
- Workbench, 3'x6' and portable toolrack, 4'x4'

7

Pests

Human-wildlife interactions are common in rural areas. Deer-car collisions are frequent, particularly around dawn and dusk in spring and fall. Wildlife may enter improperly constructed or sealed buildings and often raid trash, compost piles, gardens, and bird feeders. Attention to wildlife management can reduce unwanted interactions and allow you to enjoy the wildlife around you.

Wildlife management is often thought of in terms of protecting, enhancing, and nurturing wildlife populations and their habitat. However, many species, at one time or another, require management actions to reduce conflicts with people or with other wildlife species.

Wildlife damage control, however, is an increasingly important part of the wildlife management profession because of expanding human populations and intensified land-use practices. In addition to the need to reduce wildlife-people conflicts, public attitudes and environmental regulations are restricting use of some of the traditional tools of control such as toxicants and the types of traps that kill or maim animals.

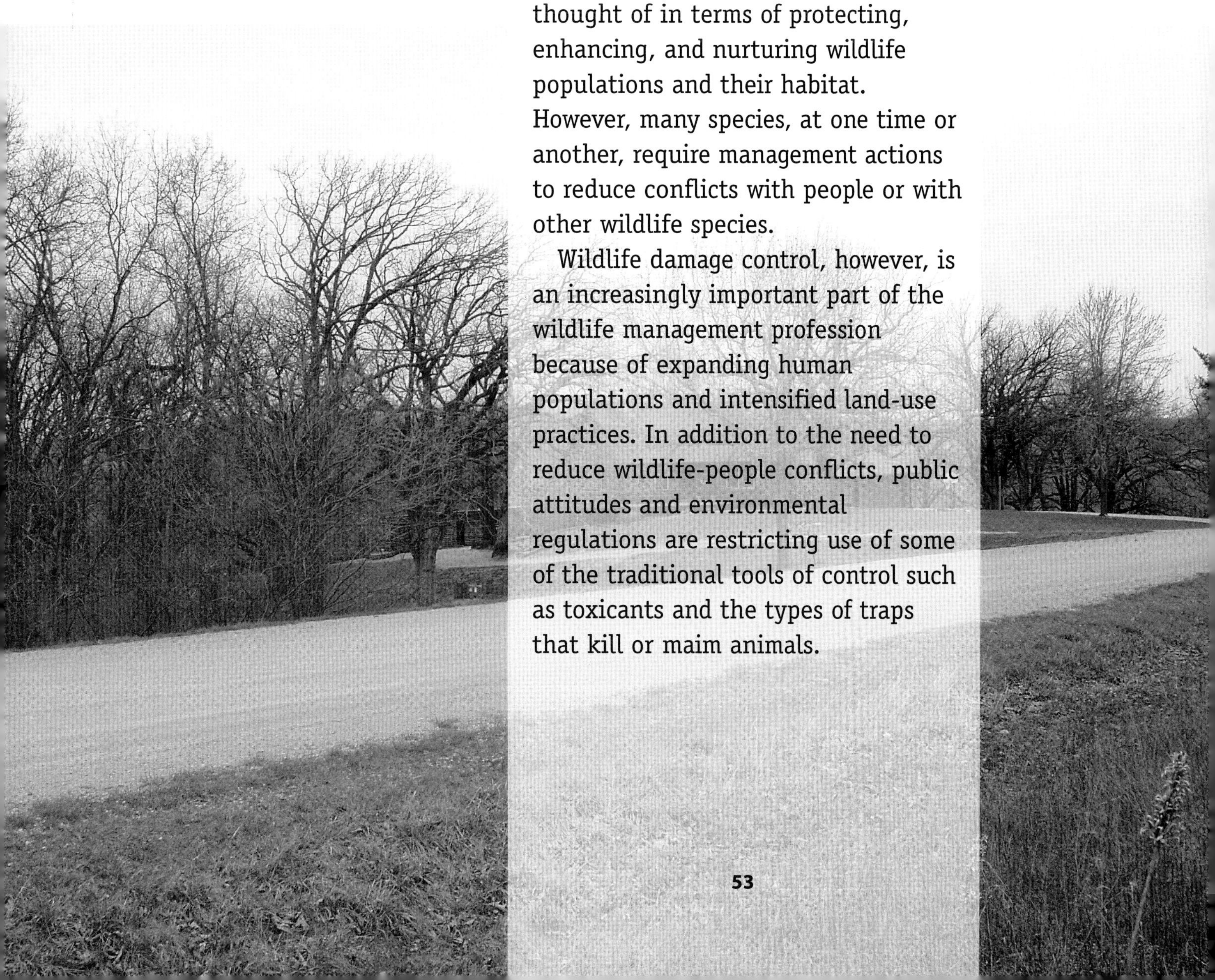

An integrated approach to damage management includes treatment of materials and techniques such as exclusion, habitat modification, repellents, frightening stimuli, toxicants, fumigants, trapping, and shooting.

Insects in the Country

People moving to the country are likely to encounter mosquitoes, horse flies, house flies, deer flies, and wood or deer ticks. House flies are a nuisance, especially if there are animals and manure around. The animals and manure can even be on a neighboring farm and still cause a problem. There are some control methods you can try, but it is very difficult to reduce the overall fly population.

In the fall, Asian Lady Beetles are a problem in many rural areas. These insects look similar to regular lady bugs, with the addition of spots on their back. Asian Lady Beetles spend their summer in soybean fields and consume aphids. After the soybeans are harvested, however, they begin looking for a place to over winter and often begin migrating to homes at that time. They often land on the south side of a home and then crawl into any opening that they can find. They can be kept out by spraying with an insecticide labeled for use inside the home. Spray around windows, doors, soffits, the corners of the house, and all around the perimeter of the house. Spray around any opening down to an eighth of an inch through which they can enter.

Mosquito control often causes frustration in rural areas, particularly if you are located near a stagnant open-water source such as a bog or swampy area. Remove all possible sources of standing water, including old tires and buckets. Keeping weeds removed from shallow pond areas, maintaining an appropriate fish population that feed on mosquito larvae, or using an aerator to keep the water circulating can reduce larvae hatching areas. It may also be possible in some areas to have the area in and around a rural home and lawn sprayed for mosquitoes by an aerial applicator. Control only lasts for about 7 to 10 days depending on the weather. The best sources of personal mosquito protection are repellents registered with the Environmental Protection Agency as pesticides. Use a spray-on repellent, which can provide protection for up to five hours, depending on the amount of perspiration, skin rubbing, temperature, and the abundance of mosquitoes. Deet-containing products are considered to be the most effective insect repellent ingredients on the market. Use products that contain concentrations of Deet at 20% to 30%, and apply the repellent to clothing rather than directly to skin.

Horse and deer flies are a nuisance, and they have a painful bite. Deer and wood ticks can't be controlled well with pesticides so be sure to check for them every night if you've been out in the woods. In most areas of the U. S., the deer tick is a carrier for Lyme disease, which can be a chronic and debilitating disease.

Insect pests, particularly heartworms, also will affect your pets. Discuss potential preventive treatments with your veterinarian.

Damage Management

Large predators have always been a concern to livestock producers. Coyotes have, at times, caused extensive damage to sheep, goats, and cattle. Red foxes, skunks, raccoons, and mink can damage waterfowl nests in small wetland areas surrounded by agricultural land.

Damage by wild animals to ornamental plants, buildings, roads, and other structures can also be serious at times. Some of the most costly problems are caused by house mice, Norway and roof rats, beavers, and deer. Wild animals also cause nuisance problems. Problems range from feces left by ducks and geese and garbage containers overturned by raccoon, to structural damage, to disturbing sounds made as small mammals move into attics and walls. Under some conditions, wild animals are reservoirs of diseases, presenting threats to other wildlife, domestic animals, and humans. People usually enjoy having wild animals near, but not in, their homes, and most are willing to tolerate some damage from wildlife.

The book *Prevention and Control of Wildlife Damage* is an excellent resource for learning about wildlife, identification of damage, solutions, and disease. You probably will not need the entire book, just the pages about the particular wildlife species causing you trouble. Look for it at your local library or bookstore. For additional information, see the Resources section at the end of this chapter, and see Chapter 6 for information about pests that can cause damage in and around farm ponds.

Wildlife Species that May Cause Problems

The following is a brief list of wildlife species and the typical damage that those species might cause.

BATS. Several bat species in the U.S. have declined in numbers during the last few decades because of chemical pesticides that have decreased the insect supply, and contaminated insects have poisoned bats. Bats can sometimes be found around swimming pools, where they can drink water and catch insects. They sometimes enter buildings through roof openings such as poorly screened areas in and around chimneys or roof vents. Bats consume vast numbers of insects, and some species look for roosting areas in attics. They can occasionally contract rabies, although few human deaths have resulted from bat-transmitted rabies. Most human exposures are the result of accidental or careless handling of grounded bats.

BIRDS. Blackbird damage to ripening corn, sunflower, and rice can be serious. Birds may depredate agricultural crops, create health hazards, and compete for limited resources with other more favorable wildlife species. Managing bird populations can often be done by the manipulation of habitats using dispersal techniques such as frightening techniques. Perhaps the greatest problem caused by blackbirds and starlings is their propensity to gather together in large, nocturnal roosting congregations, especially in winter. The noise, fecal accumulation, and general nuisance caused by large numbers of birds roosting together near human habitations can be significant. Sparrows can build bulky grass nests in buildings, drain spouts, and other sites where they can cause the backup of rain water onto roofs, fire hazards (dryer vents or chimneys) or other problems. Large raptors have been blamed for killing small animals and poultry in some areas. Pigeons and house sparrows are considered nuisances because their droppings deface buildings and cause buildings and equipment to deteriorate. Pigeons and sparrows may also spread various diseases to humans.

GEESE. Canada geese can become a problem if you have a pond on your property. They can be aggressive, especially if they have young. Droppings can also become a problem with large flocks of geese. Geese prefer open, grassy areas near or adjacent to water. Reducing or eliminating open areas near water helps reduce geese populations. Geese will damage corn crops.

CHIPMUNKS. Chipmunks damage grain storages, garden seeds, flower bulbs, and plants through burrowing and feeding. They can establish residence in, around, or under homes.

CANINES. Coyotes, wolves, and roaming dogs can prey on animals ranging from big game and livestock to rodents, wild birds, and poultry. Pets are sometimes killed by coyotes. Coyotes can also damage fruit crops. Roaming domestic dogs can be a serious problem to livestock, especially to sheep pastured near cities and suburbs.

DEER. Deer live at the forest edge and thrive in agricultural areas. They have aesthetic value but also can cause considerable damage to crops and automobiles (road accidents). Often, they harbor diseases common to humans and livestock. Deer browse (leaves, stems, and buds of woody plants) all year. They will eat corn, soybeans, small grains, alfalfa, fruits, nuts, and vegetables when available. Deer can destroy trees, shrubs, garden crops, and other landscaping. Screening or electric fencing can be used to deter damage to specific areas if damage is serious enough. If twigs appear as though they were cut cleanly at an angle, rabbit browse should be suspected. Because they have no upper incisors, deer are more likely to rip and pull bark, twigs, and leaves.

CATS. Domestic cats rarely prey on anything larger than ducks, pheasants, rabbits or quail. Feral cats are house cats living in the wild. In some rural areas, feral house cats are the most abundant predators. They are known to prey on bird populations.

FOXES. Grey and red foxes feed primarily on rabbits, hares, small rodents, poultry, birds, and insects. They also consume fruits. Foxes can kill young livestock, although poultry is their more common domestic prey.

GROUNDHOGS. Groundhogs are a significant pest, especially when they take up residence around out buildings or around pond dams. Their burrows can weaken those structures. In addition, they can consume large amounts of vegetation and damage crops.

SQUIRRELS. Ground squirrels are an important pest species in north central and western North America. They can inflict serious damage to pastures, rangelands, grain fields, vegetable gardens, and fruit or nut crops. They can carry several diseases that are transmissible to humans.

Tree squirrels eat plants and fruit, dig up newly planted bulbs and seeds, strip bark and leave from trees and shrubs and consume bird eggs. They have been known to nest in attics of homes if given the opportunity.

MICE AND RATS. Mice and Norway rats consume millions of bushels of grain each year in the field, on the farm, in the elevator, in mills, and in storage and transit. They also waste large amounts by contamination. They have been known to feed on poultry chicks and occasionally even attack adult poultry, wild birds, and newborn pigs and lambs. Viral and bacterial diseases can be transmitted to humans by rodent feces and urine that contaminate food and water.

Wood rats are attracted to food supplies left in buildings and they remove small objects such as spoons, knives, sometimes leaving sticks or other objects in trade. They are agile climbers and consume fruits, seeds, and green foliage of herbaceous and woody plants. Several subspecies of wood rats are endangered, so check local regulations before undertaking control efforts.

Rodent gnawing causes considerable property damage. Fires have been started when rats and mice gnaw through the insulation of electrical wiring. They sometimes use materials such as oily rags and matches for building nests, which can result in fires by spontaneous combustion. Rodents remove building insulation found within walls and ceilings. Control includes sealing all openings one-fourth of an inch and larger, practicing good sanitation, reducing sources of food, water, and shelter, controlling weeds, and removing debris from around structures. Anticoagulant rodenticides generally work well. Other control methods include snap traps, live traps, and glue boards. Dogs and cats are of limited value in controlling rodents.

MOLES. Moles feed primarily on soil invertebrates, especially earthworms and grubs (beetle larvae). Voles and mice use the burrows of moles and can be responsible for some damage attributed to moles. Moles can damage lawns and flower beds extensively through burrowing. Voles and mice will use mole runways to find grains, seeds, and tubers. Trapping is the most successful method of getting rid of moles.

MUSKRATS AND NUTRIA. Muskrats can cause problems where wetlands border agricultural crops. The most serious damage results from their burrows in pond dams, levees, and irrigation canals. Nutria are semi-aquatic animals that feed on aquatic plants, roots, and seeds and on crops around waterways. The greatest damage from these rodents, native to South America, is to sugar-cane and rice crops, but they have been known to damage wooden structures and floating marinas as well.

OPOSSUMS. Opossums eat fish, crustaceans, insects, mushrooms, fruits, vegetables, eggs, and carrion. They will also raid poultry houses. Habitat is diverse, ranging from arid to moist, wooded to open fields. While looking for food, opossums living near people may visit compost piles, garbage cans, and pet food dishes.

GOPHERS. Pocket gophers can cause substantial damage to agricultural crops, lawns, rangeland, and tree plantings. They feed primarily on the underground portions of plants and trees. They may also damage plastic irrigation lines as well as underground pipes, cables, and electric wires.

PORCUPINES. Porcupines are usually nocturnal and are active all year. During summer, porcupines often feed on succulent plants, including garden and truck crops in open meadows, in fields, and along the banks of streams and lakes. In winter, porcupines feed on the inner bark of trees, and girdling in the upper trunk of trees often results in dead tops. Porcupines are attracted to anything containing perspiration salt: saddles, harnesses, belts, and tool handles.

RABBITS. Rabbits eat flowers and vegetables in the spring and summer and woody plants in the fall and winter. They can damage tree plantings, gardens, ornamentals, and agricultural crops. In winter, they strip bark from and debud fruit trees, conifers, and other trees and shrubs. Control is often necessary to reduce damage, but complete extermination is not necessary, desirable, or probably even possible. They are considered game animals in most states and protected except during hunting season. Exclusion with a 2-foot high mesh fence with a 1-inch (or smaller) mesh, is an effective way to keep rabbits out of areas you wish to protect. Removing habitat (for example, brush piles, weed patches, and dumps) can help manage rabbits.

RACCOONS. Raccoons are voracious eaters; among other things, they eat mice, small birds, snakes, frogs, insects, crawfish, grass, berries, acorns, corn, melons, and cat food. Garbage cans and dumps can be a major source of food. Raccoons often damage field crops or gardens near wooded areas; they frequently eat and waste ripening corn. They raid bird nests and have, on occasion, killed small lambs. They prefer hardwood forest areas near water. Raccoons can gain entrance to attics or chimneys. The incidence of reported rabies in raccoons and other wildlife has increased dramatically over the past 30 years. In the U.S., they have been identified as a major wildlife host of rabies. Raccoons are protected furbearers in most states, with seasons established for hunting and trapping. Most states, however, allow landowners to control furbearers that are damaging their property. Exclusion, especially with electric wire, generally works well.

SKUNKS. Skunks eat plants and foods such as insects and mice. They become a nuisance when their burrowing habits interfere with humans. They can burrow under porches or buildings. Garbage or refuse left outdoors in rural areas needs to be covered securely to keep skunks out. They can also damage beehives looking for food. Occasionally, they feed on corn. Skunks are the primary carriers of rabies in the Midwest. Get professional help with their control.

SNAKES. Many kinds of snakes are found in the U.S., but only the following are poisonous: rattlesnakes, copperheads, cottonmouths, coral snakes, and sea snakes. Snakes are not very mobile, and most have specific habitat requirements. Some live underground (mostly small in size), and some live in trees (green snakes, for example). In general, they like cool, damp, dark areas where food is available. Some are attracted to firewood stacked on the ground, junk piles, flower beds with heavy mulch, gardens, unkempt basements and storage sheds, shrubbery growing against foundations, and barn lofts—especially where available feed attracts rodents, attics in houses where there is a rodent or bat problem, stream banks, ponds with debris on the banks, unmowed lawns, and abandoned lots and fields. Snakes are predators and eat many different kinds of food. Rat snakes eat primarily rodents; king snakes eat other snakes, rodents, young birds, and bird eggs. Green snakes eat mostly insects. Control by sealing small openings (with mortar, hardware cloths, sheet metal, or steel wool), reducing rodent populations, keeping vegetation trimmed and closely mowed, and removing bushes, shrubs, and other growth around buildings.

WEASELS AND MINK. Weasels and mink are carnivorous animals and feed primarily on mice, rats, moles, rabbits, chipmunks, frogs, snakes, worms, small birds, and eggs. They do occasionally feed on insects and plants. They can eat as much as 30 to 40 percent of their body weight in one night. They are tremendous predators, often times attacking animals much larger than themselves, burrowing for prey, and also swimming to capture water prey. They can create havoc in a poultry enclosure.

Disease Threats from Wildlife

Wildlife serves as a natural host for certain diseases affecting humans (zoonosis). Zoonotic diseases can be contracted from wildlife directly by bites or contamination, or indirectly through the bite of arthropod vectors such as mosquitoes, ticks, fleas, and mites that have previously fed on an infected animal. In such cases, general precautions should be taken to reduce risks of exposure and prevent infection. Protection from vector-borne diseases involves using mosquito or tick repellents, wearing special clothing, and tucking pant cuffs into socks to increase the chance of finding crawling ticks before they attach. Additional measures include checking your clothing and body and your pets for ticks and removing the ticks promptly after visiting infected sites. Reduce outdoor exposure to mosquitoes especially in early evening hours to diminish the risk of infection from mosquito-borne diseases.

Mosquito-borne Diseases

Mosquito-borne encephalitis is a disease caused by mosquito-born viruses that affect the central nervous systems. These viral infections may result in severe illness or death in humans or horses. No treatment or commercial vaccine is available for humans, but vaccines for some types of encephalitis are available for horses. The best preventive measures are personal protection against mosquito bites, avoiding exposure to mosquitoes during early evening hours, and the use of repellents. Mosquito populations can be reduced in an area by eliminating breeding sites for vector species. Killing adult mosquitoes with area wide applications of insecticides has been effective in preventing epidemics.

Tick-borne Diseases

Plague, tularemia, and leptospirosis have been acquired in the handling and skinning of rodents, rabbits, and carnivores. Humans can acquire diseases like Colorado tick fever, Rocky Mountain spotted fever, and Lyme disease if they spend time in areas where those disease vectors and hosts are found. Ticks can transmit Colorado tick fever, Rocky Mountain spotted fever, Lyme diseases, tularemia, and relapsing fever. Avoid tick-infested areas and protect against tick bites.

Bird-borne Diseases

Histoplasmosis, ornithosis, and salmonellosis are common bird-borne diseases. Precautions against acquiring fungal disease, especially histoplasmosis, should be taken when working in high-risk sites containing contaminated soil or accumulations of animal feces, especially under bird roosts or in buildings or caves containing bat colonies.

Virus-caused Diseases

The most notable virus caused disease is rabies. It is an acute disease caused by the virus rhabdovirus that can infect all warm-blooded animals, and is usually fatal, once clinical signs develop. Use caution when approaching a suspected rabid animal since many are still aggressive and can bite even if paralyzed. If the animal is still alive, it should be killed humanely without damaging the head. To confirm whether an animal is infected with rabies the animal must be submitted to the local health department or state diagnostic laboratory for testing.

Raccoons, striped skunks, and bats are the major hosts. Infected animals usually display either furious or dumb rabies, although some animals progress through both stages. Skunks, raccoons, foxes, and other canids usually have furious rabies and are unduly aggressive before convulsions and paralysis set in. Some animals, however, have dumb rabies and proceed to tremors and convulsions without agitation or aggression. Other behavioral changes include friendliness or loss of fear, appearance in daytime for some typically nocturnal species (skunks, bats), unprovoked attacks on anything that moves (including inanimate objects), bewilderment, and aimless wandering. Unusual barking, crying, and frothing at the mouth are additional signs, which are the result of paralysis of the throat muscles.

Occasionally, rabid bats are encountered prostrate or fluttering on the ground, unable to fly; they should be handled with care because they can still bite and transmit rabies. Some rabid bats, particularly solitary species, are aggressive and have been known to attack people. In domestic animals, rabies should be suspected if there is any change in normal habits, such as a sudden change in disposition, failure to eat or drink, running into objects, or paralysis.

Other Vectors

Other classes of vector-borne diseases are flea-borne diseases and rodent-borne diseases. Typical flea-borne diseases are plague and Murine typhus. Common rodent-borne diseases are rat-bite fever, leptospirosis, salmonellosis, and rickettsialpox.

Protecting Against Disease

Use extreme caution when approaching or handling a wild animal that looks sick or abnormal to guard against those diseased contracted directly from wildlife. Wear protective clothing, particularly disposable rubber or plastic gloves, when cutting up or skinning wild animal carcasses. Scrub the work area, knives, other tools, and reusable gloves with soap or detergent followed by disinfection with diluted household bleach. Avoid eating and drinking while handling or skinning animals, and wash your hands thoroughly when finished. Safely dispose of carcasses and tissues as well as any contaminated disposable items like plastic gloves. Cook meat from wild game thoroughly before eating it.

Knowledge and recognition of the early symptoms of the diseases and the conditions of exposure are essential in prevent severe illness. Know the diseases present in your general area. Contact a physician if you become sick following exposure to a wild animal. Inform the physician that such exposure has occurred.

Resources

Cooperative Extension Service offices are good places to start.

National Prairie Research Center, http://www.npwrc.usgs.gov/

Prevention and Control of Wildlife Damage, CES, University of Nebraska & Great Plains Agricultural Council Wildlife Committee & USDA Animal and Plant Health Inspection Service, Animal Damage Control.

Purdue University,

- Everything WILDLIFE, http://www.purdue.edu/wildlife
- USDA-APHIS Wildlife Services, http://www.entm.purdue.edu/wildlife/wild.htm.
- Purdue Pesticides Programs, http://www.Btny.purdue.edu/PPP/

State Wildlife and Fish Management Agencies are responsible for managing most resident species of wildlife and fish, as well as migratory species while they are within state borders. Permits are normally required before species listed as game animals, furbearers, or game fish can be controlled. Permits may also be required if species are considered rare or endangered by the state.

University of Nebraska,

- Internet Center for Wildlife Damage Management, http://www.widlifedamage.unl.edu
- UNL Extension publications, http://ianrpubs.unl.edu/wildlife/

University of Kentucky, College of Agriculture, Midwest Tree Fruit Pest Management Handbook, http://www.ca.uky.edu/agc/pubs/id/id93/ch_4.htm

USDA/APHIS Wildlife Services, http://www.aphis.usda.gov/ws/, Animal Damage Control Program, USDA-APHIS, provides operational and technical assistance to reduce conflicts between people and wildlife through the nationwide ADC program

US Fish and Wildlife Service, http://www.fws.gov/

U.S. Fish and Wildlife services have primary responsibility for managing endangered species and migratory birds.

The Electronic Zoo-Wildlife, http://www.netvet.wustl.edu/e-zoo.htm

8

Safety, Security, and Emergency Planning

Although many disasters may be unavoidable, planning and preparation can reduce your losses when they do strike. Planning for general safety, accident prevention, and emergencies is particularly important for people living in rural areas. You are more likely to experience electric outages and roads closed due to flooding or snow. Damaged electric lines and closed roads can take longer to repair in rural areas because of the greater distances involved. Furthermore, since the tax base is generally much smaller than in urban areas, there is usually a smaller response team. In fact, most people living in the country are not within earshot of warning sirens. So, it pays to be prepared and to have back-up plans. This chapter discusses basic preparedness, fire and lightning protection, security, and related topics.

Preparedness

Although many disasters may be unavoidable, planning and preparation can reduce losses. Preparation for emergencies includes the following:

- Providing areas that can serve to protect occupants from storms and tornadoes.
- Providing local emergency response organizations with your home's identity and location. Make sure your address is clearly marked on your mailbox or in another prominent location, is large enough to be seen by traffic approaching from either direction, can be read in the dark, and is not obscured by vegetation.
- Installing and maintaining communication/warning systems (replace batteries and test at least annually).
- Installing and maintaining a standby electric power system.
- Having plans for exiting buildings.
- Installing fire extinguishers and other fire control schemes.
- Keeping first aid, food and water supplies, emergency lighting (flashlights, candles, etc.) and a battery-powered radio readily available.
- Maintaining first aid skills.
- Inspecting safety systems regularly to minimize malfunctions.
- Providing a prioritized list of contact names and phone numbers that should be notified in case of an emergency if you are not available.
- Storing important papers in a locked, fire-proof cabinet or in a bank safety deposit box. Be aware that most home fire-proof cabinets will not protect computer discs, which can be damaged at much lower temperatures than paper documents.

The normal human response to the destruction of facilities in a disaster is to rebuild them just as they were before and to do so as quickly as possible. While dealing with the disaster often requires immediate attention (e.g., putting out the fire), long-lived decisions (such as rebuilding) are often better postponed until damage is fully assessed, family goals and needs are thoughtfully reevaluated, and alternatives are carefully examined. Delay also allows for a better assessment of insurance assistance and the owner's overall financial situation. Recovering from a disaster can sometimes provide an opportunity for making major changes.

Insurance

Keeping insurance up-to-date is as important in rural areas as it is in urban areas. Make annual appointments with your insurance agent to review your insurance coverage. It is important to discuss specific exclusions and amendments in your coverage. Get any agreements in writing. Make sure you and the agent fully understand the limits of coverage. Ask if installing security measures around the home will lower your premiums.

Most insurance companies have affordable coverage available for more common disaster situations, but special coverage may be needed for specific facilities, such as large livestock operations or specialty buildings. Check to see if your insurance covers damage caused by perils such as lightning, wind, snow, fire, and flood. Be sure the policy provides personal liability insurance.

If you have a business that brings visitors or customers to the farm (such as a u-pick fruit operation or a bed and breakfast), make sure that you have liability insurance that adequately covers that operation.

Signage

You should be familiar with common safety signage in rural areas and ensure that you and your family are aware of potential dangers. Figure 8-1 shows typical examples of several common safety signs. The example in the upper left corner of the figure shows the model commonly used for complex signs such as the toxic chemical hazard and high-voltage signs in the figure.

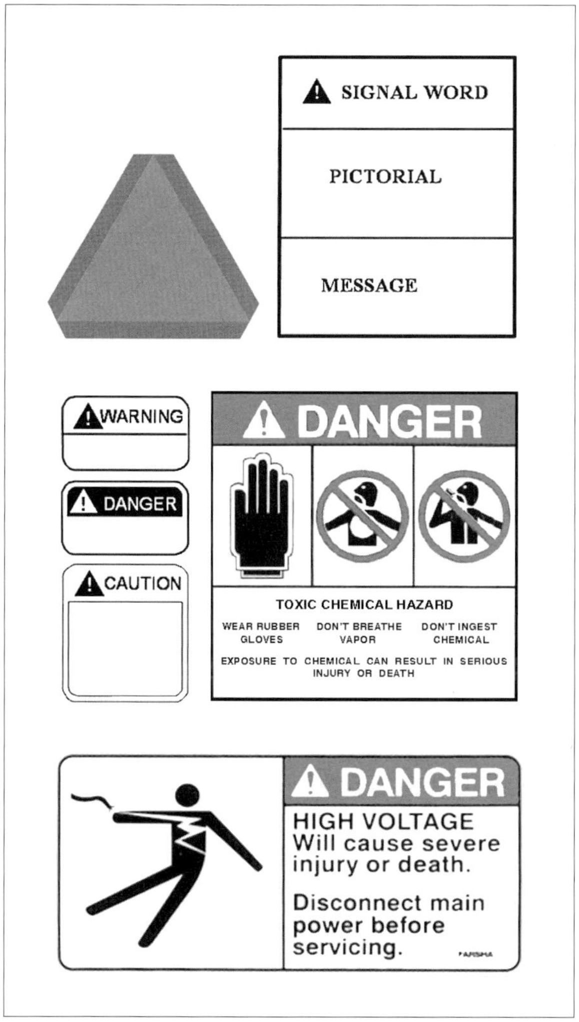

Figure 8-1. Typical safety signs in rural areas.

Fire Prevention and Protection

Fire prevention in rural areas is very important because rural fire departments usually depend on volunteers and because of the distance from homes to the fire station. Consequently, the response time is usually greater than in urban areas with full-time firefighters. Preventing loss from fire requires good housekeeping, properly maintained heating equipment, water for fighting fires, lightning protection, and effective warning systems. Strategically placed smoke detectors can give an early warning that is critical. They should be located on each floor of the home, including the attic, because smoke rises. Approved, portable, regularly maintained fire extinguishers should be located near high risk locations such as furnaces, fuel storage facilities, vehicle parking areas, and shop areas. Fire extinguishers need to be kept filled and regularly checked. As an added element of safety planning, make sure the local fire department is familiar with your location, layout, and availability of water.

Public water supplies can be limited in rural areas so ready access to a suitable private water supply is important. A water system capacity, minimum 10-gallons per minute (gpm), is necessary for fire control. Fire fighting water can be supplied from a large swimming pool or nearby farm pond, stream, or lake using PTO or engine-powered pumping equipment. Electric power may fail or be disconnected during a fire at a rural site, so a backup power supply that doesn't depend on the electric utility and on pumps and hoses is an excellent safety measure.

To reduce fire hazards:

- Use fire-resistant building materials.
- Locate water hydrants (on 1.25 inch diameter water line minimum) adjacent to buildings.
- Locate chemical fire extinguishers adjacent to grain-drying equipment, welders, and fuel storage facilities. Keep extinguishers handy in buildings and vehicles.
- Space buildings adequately to prevent rapid fire spread. Fire codes usually requires a minimum of 10 feet between the eaves of neighboring buildings
- Install smoke and carbon monoxide alarms.
- Keep grass and debris around buildings mowed and removed.
- Do not use burn piles and burn barrels within 50 feet of buildings and combustible materials.

A swimming pool, cistern, pond, or other water source can provide emergency water. A rural pond can be an excellent source of water for firefighting. Although most rural fire departments are equipped to draw water directly from the surface of a pond, if you build a pond, it is best to construct a dry hydrant in the pond embankment at the time of construction. Figure 8-2 shows this type of installation.

A dry hydrant allows water to be pumped through a permanently installed screen near the bottom of the pond, up through an adequately sized pipe to a capped hydrant on top of the pond bank. In northern climates, the pipe must be buried below the frost line so it does not freeze in winter. Avoid a suction lift of more than 20 feet since this makes it more difficult to begin the pumping process. Provide an all-weather road access to the dry hydrant, and keep snow removed in winter. Work with your local fire department to develop a regular system of inspection, maintenance, and testing for the hydrant. Details on installing and maintaining a dry hydrant can usually be obtained from rural firefighting handbooks. See MWPS-2, *Farmstead Planning*, from MidWest Plan Service for information on maximum frost penetration for various areas around the U.S., or check with local excavators.

Field firebreak strips can be used to protect rural homes from fast-traveling, open-field type fires if you are located in an area that is prone to grass fires. Generally, such firebreak construction surrounds a rural living site and involves a space of tilled soil at least 100 feet wide and/or several tilled strips with backfiring (a small, controlled secondary fire to reduce the amount of combustible material available) done between them. Range management or forestry agencies may have unique firebreak construction recommendations for your area that may include grass, brush, and/or forest fires. Local firebreak planning and construction is a part of planning for your rural home site and needs implementation long before a fire approaches.

Lightning Protection

Lightning is a frequent threat to property and personal safety in rural areas. Damage to electric wiring and equipment and house fires can also occur. For personal safety during a lightning storm, stay indoors or in a rubber-tired vehicle, and avoid the use of corded telephones. A high-voltage surge from outdoor, overhead electric wires can result from a direct or even near miss lightning strike, damaging electronic equipment, or causing fire. Consult with your local power supplier about using a lightning surge arrestor at the electric service entrance panel to

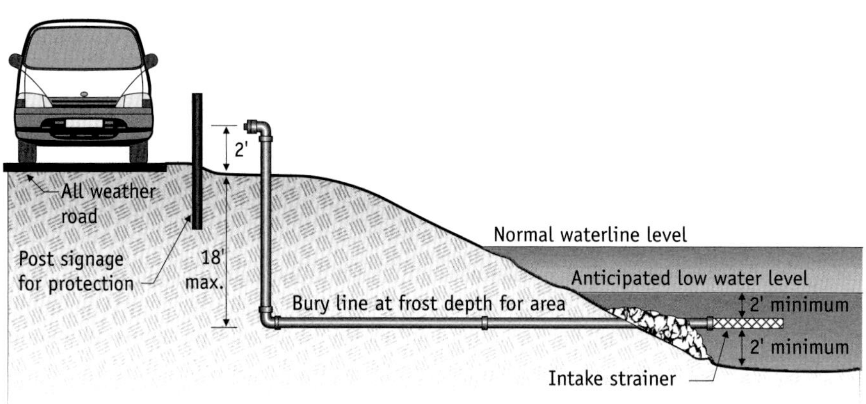

Figure 8-2. Dry hydrant construction.
Adapted from *Fire Protection in Rural Areas: Dry Hydrants for Ponds.* The Ohio State University Fact-Sheet AEX-422-98. (http://ohioline.osu.edu/aex-fact/0422.html)

reduce the voltage surge and conduct it to ground via a ground rod. In most states, lightning protection reduces the cost of fire insurance on buildings, and a portion of the money spent on lightning protection may be recoverable on your homeowner insurance premiums. Also, your building insurer may have specific recommendations about lightning protection needs and installation details in your area, so consult with your insurer before installing a lightning protection system.

Lightning rod protection on a building exterior that meets the requirements of the Lightning Protection Institute should conduct voltages from lightning strikes harmlessly to the ground via wire conduits. A properly designed, installed, and maintained lightning protection system affords almost 100 percent protection to a building. Lightning protection systems should be installed only by persons who have the necessary training and equipment. For the most effective lightning protection system, obtain a Master Label System approved by the Underwriters Laboratories, Inc.

A lightning bolt generally prefers to follow a metallic path to ground. At points along the path, the main bolt may leap to metal plumbing, or parts of the current may jump to appliances, metal water lines, or even to people or animals. Lightning may strike trees, buildings, or other objects because such materials are better electrical conductors (easier paths to ground) than air. Lone trees and isolated rural buildings are prime lightning targets because they provide a "ladder" for ground charges to climb nearer to the cloud charges.

Protect all important buildings from lightning damage. Minor buildings, trees, and wire fences also may require protection. Lightning can enter a building in four ways:

1. Striking a building directly.
2. Striking a metal object, such as a television antenna or cupola.
3. Striking a nearby tree and leaping to a building to find a better path to ground.
4. Striking and following a power line or an ungrounded wire fence to a building.

Lightning protection systems are designed to provide a direct, easy path for the lightening to follow to ground and to prevent damage, injury, or death as the bolt travels that path. All points where a bolt is likely to strike and all objects to which the current might sideflash must be protected.

Lightning protection of homes consists of seven principal considerations: air terminals (lightning rods), main conductors, secondary conductors, lightning arrestors, ground connections or rods, metal clad buildings, fences, and trees. They are described below.

AIR TERMINALS. Air terminals are pointed metal rods or tubes installed on high points and other strategic areas of a building. Figures 8-3 shows recommended lightning protection for houses. The air terminals attract lightning bolts that strike in the immediate area. The contour and slope of the roof determine the length of a terminal, but, terminals are usually 10 to 24 inches high. The shorter lengths are less conspicuous. All chimneys, dormers, ventilators, flagpoles, towers, water tanks, and other projections need at least one terminal. Silos and towers with flat roofs require two or more.

MAIN CONDUCTORS. The main conductors connect air terminals with each other and with grounds. Their purpose is to rapidly conduct the lightning bolt safely from the point struck to the ground. All secondary conductors are also connected to the main conductor. Each building requires at least two down conductors to the ground for each air terminal, (Exception: Conductor drops from a higher to a lower roof level are permitted without an extra downlead, provided there are no more than two ridge or three flat roof air terminals on the lower level).

SECONDARY CONDUCTORS. Secondary conductors connect metal parts of a building together to provide an electrical bond to keep lightning from flashing across the gap between the metal parts. These parts are divided into bodies of conductance and bodies of resistance.

ARRESTORS. Arrestors prevent dangerous surges of electricity into a building's wiring system when lightning strikes the power

Figure 8-3. Lightning protection for houses.
(**1**) Space terminals no more than 20' o.c. along ridges and within 2' of ridge ends; (**2**) install at least two downlead conductors with (**3**) grounds, at least 10′ deep, for house—need additional grounds for clotheslines, etc.; (**4**) tie roof projections such as ornaments into conductor system; (**5**) protect trees within 10' of house—connect to house grounding; (**6**) install at least two terminals on chimneys; (**7**) install terminals on dormers; (**8**) install an arrestor on antenna—connect to main conductor; (**9**) tie in gutters; and (**10**) make sure there is a lightning arrestor on overhead power lines.

lines. A power surge caused by lightning occurs so rapidly that ordinary circuit breakers and fuses cannot protect wiring and electrical equipment. Arrestors divert power from the hot wire and safely guide it to the ground when the electrical charge on the wire exceeds a given level.

GROUND CONNECTIONS. Proper grounds are critical to ensure dissipation of a lightning discharge without damage. The required extent of grounding depends on the soil, ranging from two simple, 10-foot grounds for a small building located on deep conductive soil, to an elaborate network of cables and rods or plates buried in dry or rocky soil that has poor natural conductivity. Figure 8-4 depicts proper construction of a ground connection.

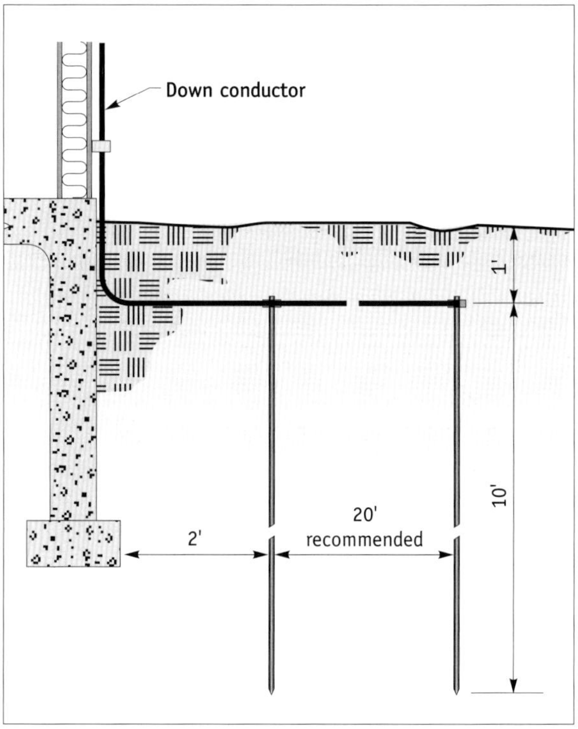

Figure 8-4. Ground rods.

Ground each down conductor separately. Buildings with perimeters of more than 200 feet require one additional down conductor with ground for each additional 100 feet of perimeter or fraction thereof. Separate the down conductors as much as possible; place them on diagonal corners of rectangular buildings and on diametrically opposite sides of cylindrical structures.

In most clay soils, grounding is accomplished by clamping the conductor cable to a ground rod driven at least 10 feet into the ground. For sandy soils, connect two or three ground rods in parallel. Space them at least 6 feet apart with a preferred distance of 20 feet and drive them 10 feet deep. Make connections between ground rods and conducting cables at least 2 feet from the foundation wall and 2 feet below ground level. This prevents damage to the foundation from pressures created by lightning dissipating in the ground.

Secondary service arrestors and arrestors for television antennae lead-in wires are recommended for more protection. Water pumps, motor frames, and enclosures for electrical equipment must be properly grounded. Contact your local power supplier or local inspection authority for details or refer to the NEC Article 250. The earth must not be used as the sole grounding conductor for equipment (NEC 250-51 and 250-91c). Submersible water pumps and associated metal well casings also must be grounded (NEC 250-43(k) and 250-43(1). Reduce pump damage from lightning with proper installation of approved lightning arrestors at the pump control center. Consider using pumps with built-in lightning arrestors.

METAL CLAD BUILDINGS. When using a building's steel framework as the main conductor, connect air terminals directly to the framework or to conductors leading through the ridge to the framework. Make the connections on cleaned areas of the framework. The surface contact area must be at least 8 square inches. Connections can be bolts, welds, or bracings.

TREES. Trees struck by lightning pose a threat of falling on buildings, equipment, or people. Therefore, lightning protection is important for valuable trees and for large trees within 10 feet of any building. To protect trees, install an air terminal at the top of the trunk (or main branch) and at the end of main branches (Figure 8-5). The main conductor of the terminal should run from the trunk to the ground. Connect terminals located on branches to the main conductor. Locate air terminals at the outreaches of main branches. Locate the grounding rod away from the trunk at branch line, but at least 12 feet away from the trunk. Locate the grounding rod at least 12 feet away from the tree. When the reach of the branches is greater than 12 feet from the tree and locate the grounding rod at the edge of the reach of the branches to prevent damage to the root system. Drive a 10-foot ground rod vertically into the ground out of reach of the tree root system. Keep the conductor away from the root system. Bury the conductor from the tree to the grounding rod in a shallow trench so livestock or equipment will not damage the conductor. Cover the conductor/ground rod if needed to prevent damage.

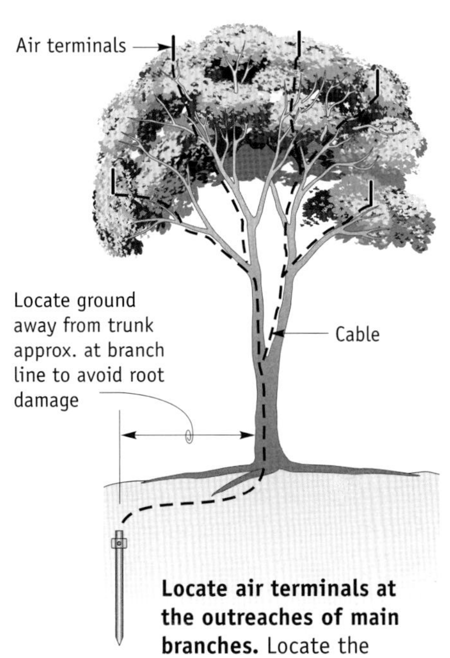

Figure 8-5. Grounding of valuable trees or ones that could cause damage to property or livestock.

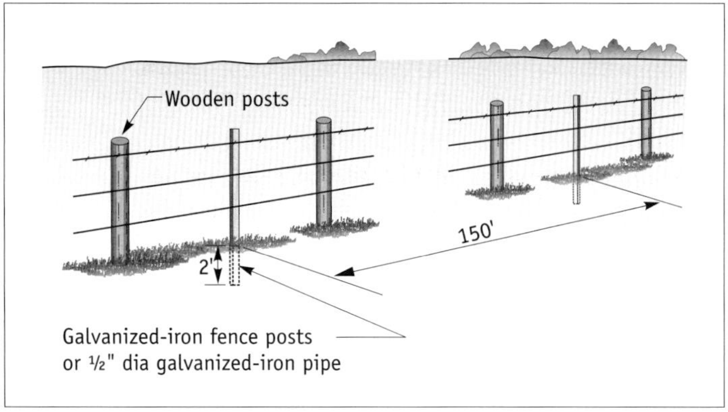

Figure 8-6. Grounding wire fences.

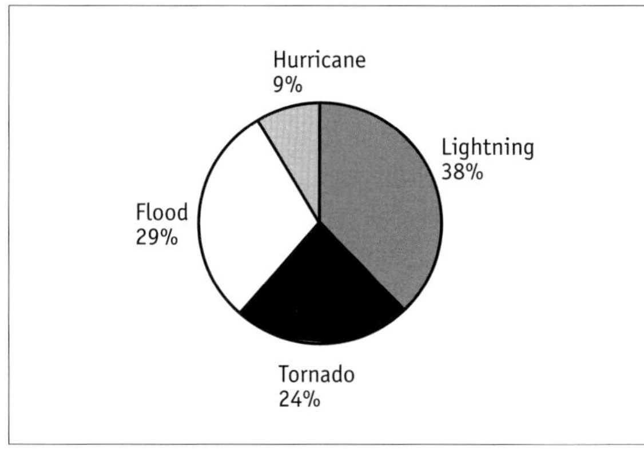

Figure 8-7. Weather-related deaths 1940-1989.
Based on National Weather Service (NWS) data.

FENCES. Improperly grounded wire fences that have been struck by lightning may carry current along the wires for up to 2 miles. Ground fences that are attached to trees or buildings, those with wooden posts and those with steel posts set in concrete. Wire fences can be grounded in one of two ways:

- Inserting galvanized iron fence posts at 150 feet on-center along the fence line (Figure 8-6).
- Driving a galvanized iron pipe beside the wooden fence post and attaching the fence wires with wire ties. Space pipes at a distance of 150 feet on-center. Use a pipe diameter of at least 1/2 inch and drive the pipe into the ground at least 2 feet.

Floods and Wind Storms

Wind storms and tornadoes are relatively common during spring and early summer in the mid-section of the U.S., and hurricanes are common in southern and coastal areas of the country at other times of the year. Wind storms often are followed by a heavy rain or thunderstorms that can cause flooding. Figure 8-7 shows the average number of storm-related deaths in the U.S. according to the National Weather Service. Wind storms can involve sustained, straight-line winds of more than 75 miles per hour (mph) with gusts greater than 100 mph. Major damage often includes broken and uprooted trees, downed overhead power and utility lines, blown-down buildings, blown-off roofing, overturned shelters, and damaged equipment. Likewise snow storms are common in many northern climates. Tall tree shelterbelts and buried utility lines help minimize damage, and good housekeeping can reduce debris scattered by storms.

Floods

Floods, such as that seen in Figure 8-8, are a natural, inevitable part of rural life in many areas. Spring flooding may be caused by rapid snowmelt and/or sustained, heavy rains and are relatively predictable. More unpredictable are sudden, heavy rains that cause local flash flooding. Information about the incidence of local flooding should be available via the County Emergency Planning office, local newspapers, and long-term climate records. FEMA has some information and publications available, such as *Design Manual for Retrofitting Flood-Prone Residential Structures* that can be downloaded or purchased from their website (http://www.fema.gov).

If you plan to build, locate your house above the 100-year flood plain to minimize the possibility of flooding. Local zoning offices often have maps showing the 100-year flood plain. Many localities or states have building restrictions to ensure that building does not occur in a flood plain.

Floods can damage trees and buildings. A 2-foot depth of flowing water can float and carry away most vehicles, and receding floodwaters leave behind saturated soils, debris, and weed seeds, and softened soil

Figure 8-8. Flooded farmstead.

plus high winds can result in downed trees and failure of buildings with shallow foundations, including post frame buildings. Protect water supply wells from floodwater contamination. Wave action caused by high winds during flooding can quickly destroy the sidewalls of flooded buildings. Properly placed tree shelterbelts can reduce such losses.

Wind Storms

Tornadoes are local storms of short duration formed by winds rotating at very high speeds, usually in a counter-clockwise direction. Tornadoes usually form during warm, humid, unsettled weather in conjunction with a severe thunderstorm. The forward speed of tornadoes ranges from almost zero to 70-mph, averaging about 40-mph. The average path is 1/4-mile wide, and tornadoes seldom travel more than 16 miles, but spectacular incidents have been a mile wide and 300 miles long. The Midwest and Plains states and the Gulf Coast are a favorable location for tornadoes during April, May, and June; however, tornadoes can occur anytime during the year. Hurricanes occur in southern and eastern coastal areas from mid summer until the end of November.

If you live in an area with frequent, severe windstorms you may decide to build an underground shelter. If you do, be sure to locate it in a well-drained place. Refer to the Resources section of this chapter for publications with more information. Aboveground safe rooms also can be used. In the absence of a storm shelter, go to a basement in the event of a tornado. In hurricane-prone areas, follow the evacuation orders of local officials.

A storm shelter constructed of reinforced concrete (Figure 8-9) provides reliable family protection against a windstorm or tornado and may be justifiable in areas of high storm prevalence. Prefabricated shelters also are commercially available. A location near the family living area is convenient and simplifies maintenance. The FEMA publication, *Taking Shelter From the Storm*, FEMA-320. (http://www.fema.gov/pdf/fima/fema320.pdf) provides information on building a safe room in your house.

Winter Storms

Studies of wind speeds and snow accumulations have led to the development of general guidelines for planning rural homes and farmsteads to counteract the effects of wind and snow. In the North Central United States, tree windbreaks (or shelterbelts) are used to reduce wind velocities and collect most wind-driven snow before it reaches the protected area. Constructed windbreak fences are often placed around parking and other areas needing wind and snow protection.

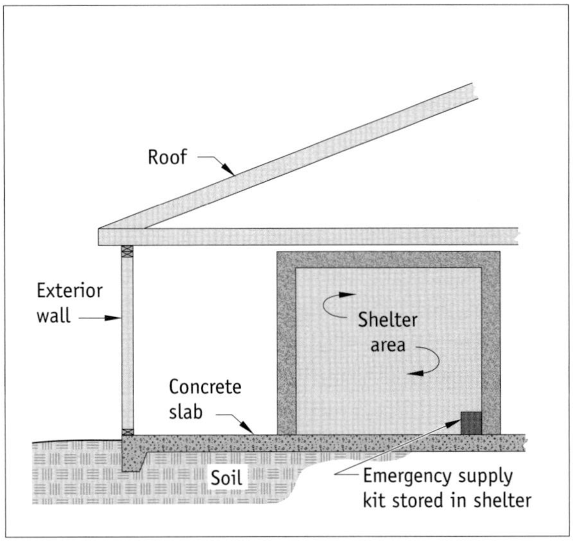

Figure 8-9. Storm shelter, below ground, reinforced concrete.

Blizzards and Snowfalls

The U.S. Weather Service defines a blizzard as a storm with winds of more than 35 mph accompanied by considerable snow and temperatures below 20° F for an extended time.

Blizzards often create large snow drifts on the leeward side of roofs. Large snow drifts on one side of the roof can create a load imbalance that can lead to roof collapse. Buildings located too near windbreaks, and old buildings are especially vulnerable to collapse failure. Reduce problems with snow loads on roofs by properly locating windbreaks with respect to buildings and by designing and constructing buildings to meet snow/wind load building construction codes for the area. Regular inspection during winter and removal of snow accumulations over 15 to 20 inches deep can help prevent roof collapse. Buildings surviving heavy snow loads can be weakened so even a small snow load in later years can cause a collapse. Have buildings that have received a heavy snow load inspected by a knowledgeable professional. Check the building for movement at connections and changes in alignment of structural members (that is, the roofline), and repair and reinforce as needed.

In northern climates, snow or icicles avalanching off a poorly insulated roof can result in damaged roof gutters and injuries if it occurs near an entrance. As heat from the building moves through the roof, the snow nearest the roof surface begins to melt, providing a lubricated surface on which snow buildup can slide relatively easily when jarred (for example: by the opening of a door). Snow jacks installed on roofs help hold the snow on the roof while allowing rain and snowmelt to drain. The can minimize the snow avalanche potential. Likewise, roof gutters can reduce icicle formation. The areas around large doors, walk-through doors, and walkways along the sides of a building are the most hazardous places where snow avalanches occur. Locate the main personnel entrance on the gable end of the building or consider protecting the entry with a covered area if you live in an area with heavy snows.

Preparedness

When wind speed is reduced, snow, dust, leaves, and other wind blown materials settle out of the air stream. Therefore, efforts to control wind must allow space for snowdrifts and other deposits to build up where they do not cause problems. Windbreaks are most effective when placed at right angles to prevailing winter winds. Although wind may blow predominantly from one direction during winter, it rarely blows exclusively from that direction. As a result, protection is not equal for all areas on the leeward side of the windbreak.

Wind rose diagrams, such as those displayed in Figure 8-10 indicate average observed wind speed from each direction. The number in the middle of the rose indicates the percent of time there is no noticeable wind, and the 16 spikes emanating from this center circle indicate the percent of time that wind blows from that direction during the month. The wind rose does not give any indication of wind speed.

In Figure 8-11, for example, in Des Moines, Iowa, during January, about 7% of the time the wind blows from due south, and about 8% from the north, with 13% from the northwest. The July map (Figure 8-12), on the other hand, shows the wind blowing more from southerly directions. This information can be used by a family choosing a home site.

Guidelines to minimize snowdrift problems in areas of heavy snowfall:

- Use trees, shrubs, fences, and other devices to catch drifting snow before it reaches the protected areas or buildings.
- Place drive lanes and roads parallel to—not perpendicular to—prevailing winds. If this is not possible, elevate the roadway at least 12 inches above the surrounding terrain.
- Avoid locating roads directly upwind or downwind of large obstructions; maintain a 100-foot clearance upwind and a 200-foot clearance downwind.
- Pile removed snow from parking areas downwind, not upwind of the lot. Consider snowmelt drainage when locating piled snow.

- Provide adequate space for snow removal and storage between large buildings if the buildings are oriented perpendicular to the prevailing winter wind.

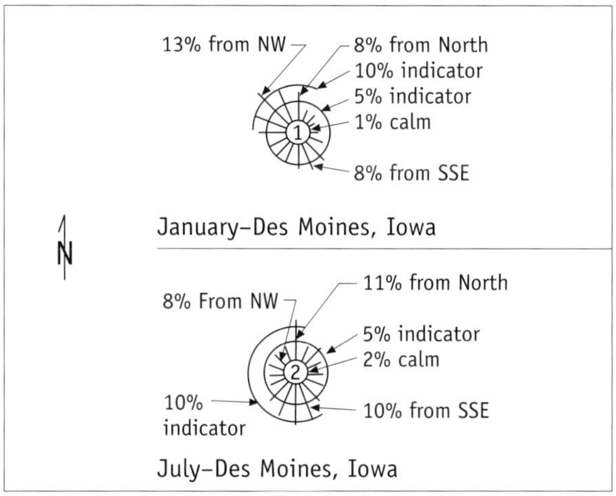

Figure 8-10. Interpretation of a wind rose.

- Locate building doorways to minimize the effects of wind and blowing snow. Doorways on the upwind end are difficult to seal; therefore, it is usually better to locate doors on the sides of building or on the downwind end.

Shelterbelts

Shelterbelts are strategically planted rows of trees and shrubs used to diminish wind effects on humans, livestock, plants, and property. They provide many benefits:

- Reduced snowdrift problems.
- Area beautification and privacy with color and landscape form.
- Reduced noise levels from nearby vehicles and other sources.

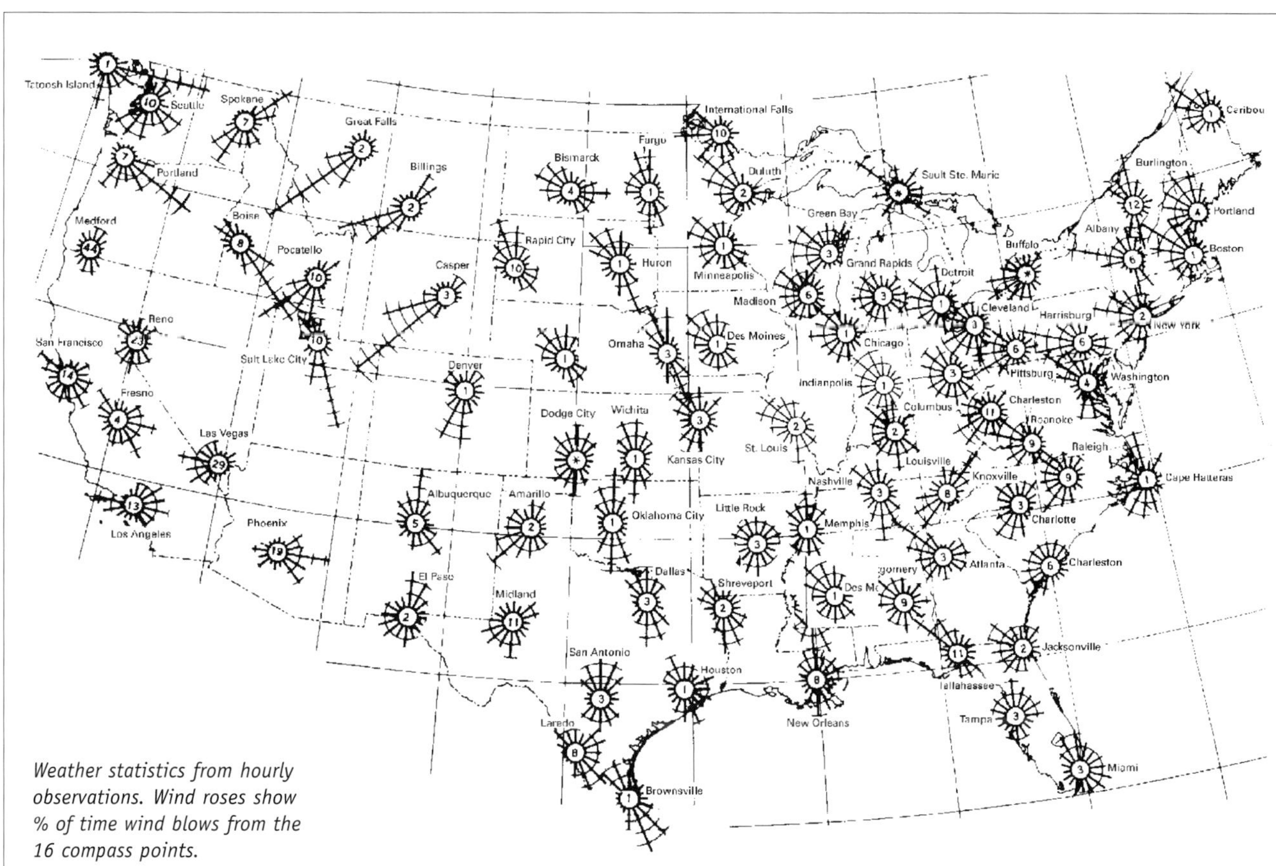

Figure 8-11. Typical surface wind roses for the U.S. for January.

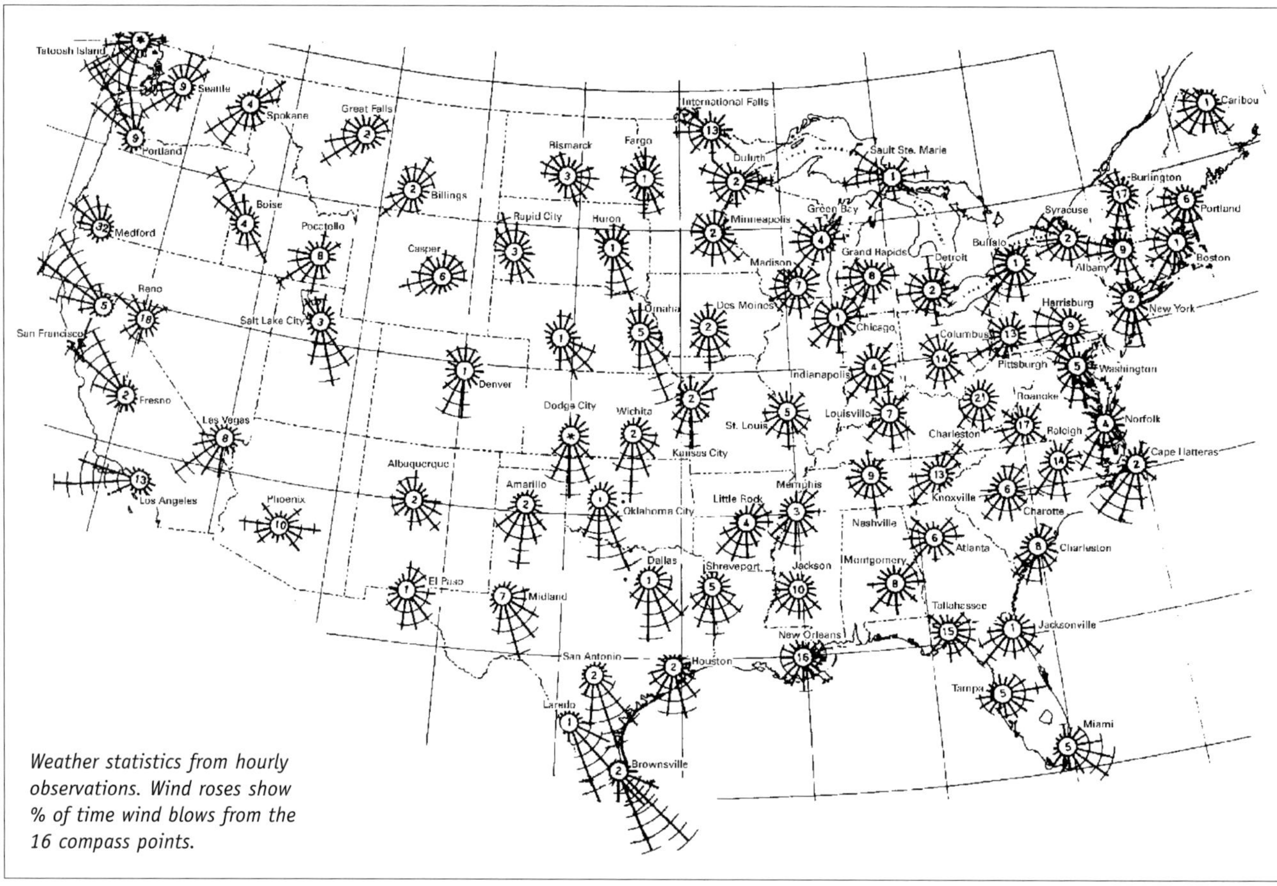

Weather statistics from hourly observations. Wind roses show % of time wind blows from the 16 compass points.

Figure 8-12. Typical surface wind roses for the U.S. for July.

- Reduced energy use for heating and cooling buildings.
- Improved outdoor working conditions and odor control.
- Reduced wind damage to property.
- Wildlife habitat.
- Filtering out blowing soil, dust, and other debris from nearby fields.
- Potential provision of fruit, nuts, and firewood.

The major steps in shelterbelt design include:

1. Locate the shelterbelt: Consider prevailing winds, site conditions, soil type, property lines, and other potential land uses. Determine width and length of shelterbelt needed. A shelterbelt width of 2 to 3 rows is the minimum; 10 rows is the maximum needed, with adequate row length to give protection from expected wind changes and enough shelterbelt width to provide adequate density throughout the height of the shelterbelt.

2. Select species: Choose plantings that can adapt to the site, and provide adequate density at the lower, middle, and upper levels of the shelterbelt.

3. Arrange the shelterbelt: Put shrubs, short trees, and slower growing trees in outer rows to provide adequate light and space between fast and slow growing trees to avoid suppressing growth. Consider tree appearance in outer rows.

4. Determine tree spacing: Between-the-row spacing of 15 feet to 30 feet and in-row spacing between plants of 3 feet for shrubs to 20 feet for tall trees are general recommendations. Make sure there is enough space between mature rows of trees for tractor cultivation and maintenance equipment.

Tree and shrub variety selection for a shelterbelt must be appropriate to the climate and soil conditions. Each species has its own characteristics of winter hardiness, growth rate, height, width, density, longevity, and resistance to insects and disease. Local nurseries and the USDA Natural Resource Conservation Service office should have recommendations for your area. Observe the strengths and weaknesses of older trees in your area. Consider using trees that will produce firewood, lumber, nuts, fruit, or wild life benefits.

Although shelterbelt height determines the extent of leeward wind area protected, shelterbelt length determines the total protected area. Shelterbelts including tall coniferous and deciduous trees reduce wind velocities perpendicular to their side for 5 to 10 tree heights upwind and 10 to 30 tree heights downwind. The protected area is smaller if the wind strikes the shelterbelt at an angle. For extreme wind and snow areas, include tall and short trees, evergreens, and shrubs in the shelterbelt. The width of the reduced-wind area and the width of the snow catch area increases as the shelterbelt trees grow. For added snow catch space, consider additional tree planting upwind (Figure 8-13.).

Locating buildings, livestock feeding areas, and service roads at least 125 feet downwind from the downwind side of a shelterbelt minimizes snowdrift problems in these areas (Figures 8-14, 8-15, and 8-16).

Shelterbelts can be effective in reducing energy needs for building heating/cooling by reducing wind and building air-exchange rates. Savings from shelterbelts can reduce annual home heating costs by up to 10-15% in the Northeastern U.S. and 15-25% in the windy North Central U.S. Shelterbelts are cost effective in these areas, based on the value of heating energy savings exceeding costs for shelterbelt establishment and maintenance over a 20-year shelterbelt life.

Figure 8-15 shows plantings around the perimeter of a home. Shrubs/conifers planted

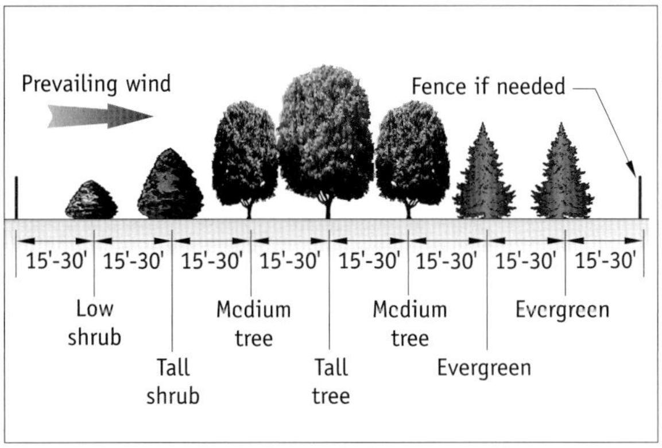

Figure 8-13. Typical shelterbelt tree spacing.

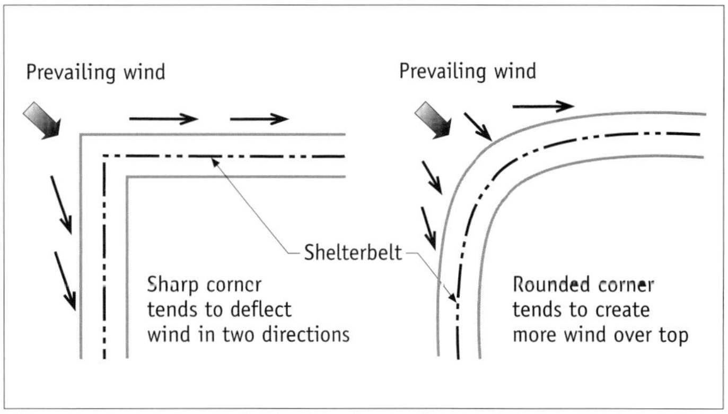

Figure 8-14. Tree shelterbelt orientation and farmstead.

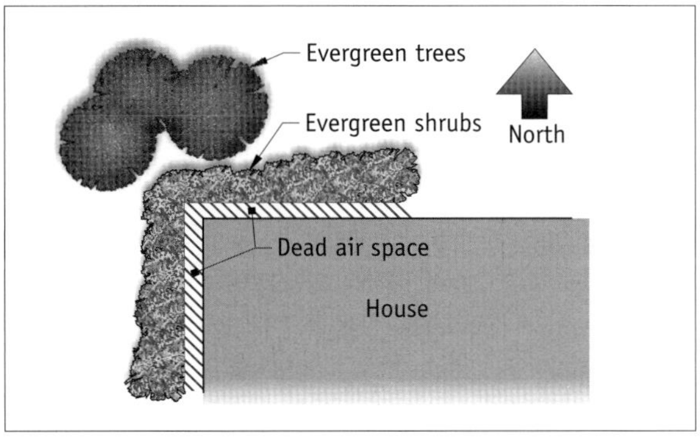

Figure 8-15. Shrubs/conifers planted around the perimeter of a home.

Figure 8-16. Shelterbelt location effect on wind and snowdrift.

around the perimeter of a home reduce energy loss in winter. Take care where they are planted as too wide a space between plantings and the building can cause a wind tunnel effect and actually increase wind speed in the area.

Windbreak Fences

Constructed windbreak fence barriers are used for localized wind and snowdrift control. The actual protected area depends on wind speed and direction, barrier height, and porosity. A completely solid barrier provides snow protection for only a short distance downwind (1 or 2 barrier heights). A solid windbreak fence is useful for increasing protection for small areas that already have general wind protection.

Compared to a solid windbreak fence barrier, a partly-open or porous, 80%-solid fence reduces wind speed for a greater distance downwind and spreads out drifted snow. A porous windbreak fence is useful for protecting larger areas, Figure 8-17 shows a comparison of the effects created by using a solid fence as opposed to a porous fence.

Various configurations of windbreak fence barriers are effective. A 10-foot high, part-open or slotted windbreak fence on a 4-foot high, built-up, earthen ridge (for improved drainage) may provide wind protection for 200 feet downwind (Figure 8-18). Note, however, that a wind from a direction different than the barrier was designed for may actually increase snowdrift or draft problems. Windbreak fences combined with shelterbelts can improve the deposition patterns of snow and soil (Figure 8-19).

General Security Issues

Security can be a difficult problem for isolated homes. They are vulnerable to vandalism, theft, arson, and bioterrorism. Vandals and burglars often take advantage of

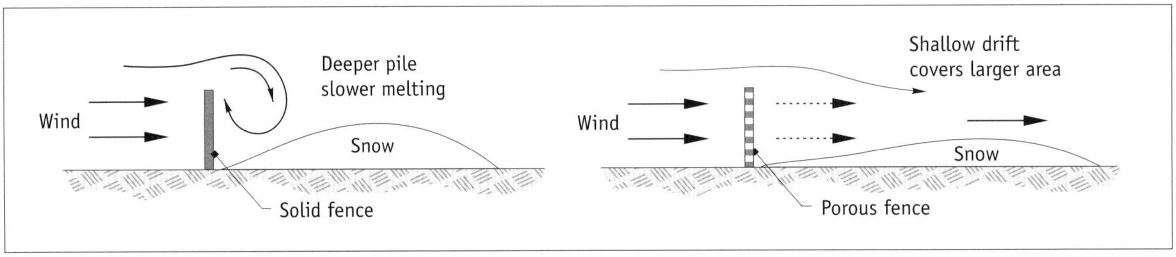

Figure 8-17. Wind protection from solid and porous fences.

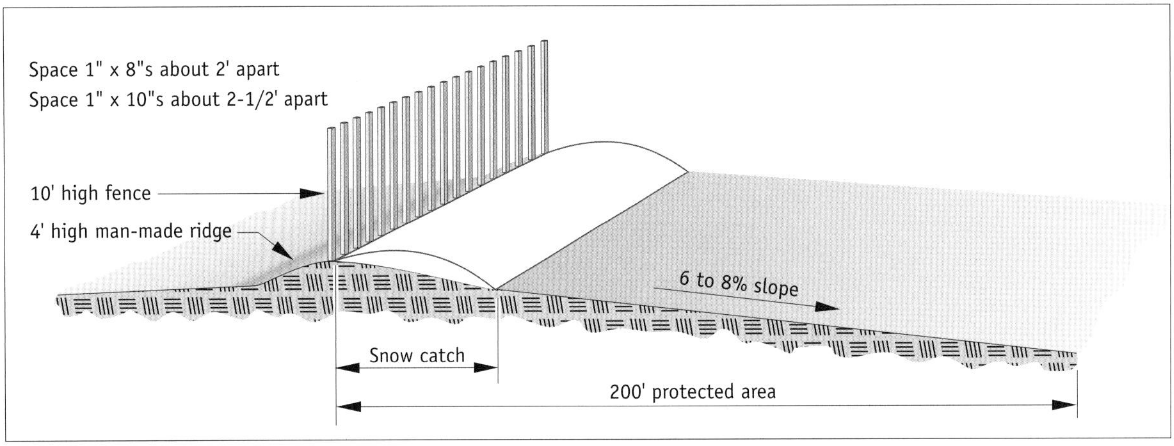

Figure 8-18. Porous windbreak fence.

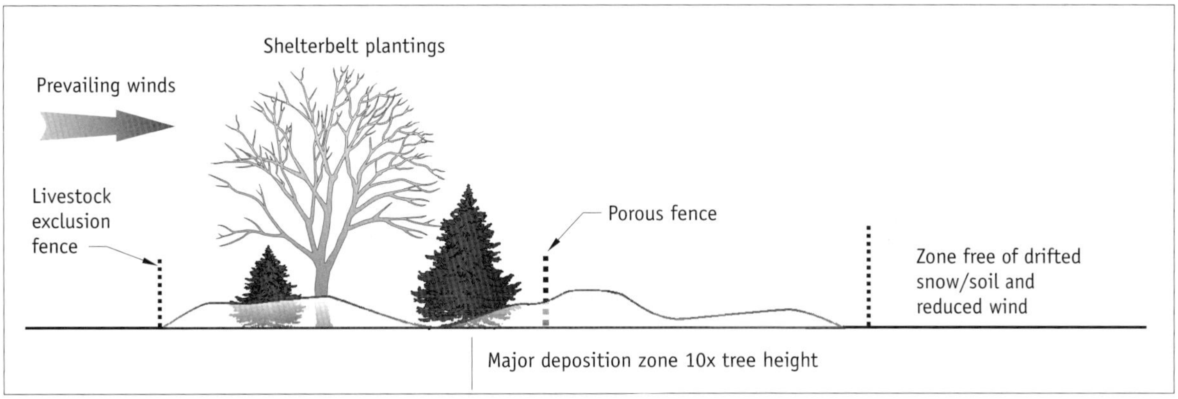

Figure 8-19. The combination of fence and shelterbelt plantings greatly enhances the pattern of snow and soil deposition.

sparsely populated communities with minimum security and police protection, poor lighting, isolated buildings, and valuable equipment parked outdoors. Homes can be protected by use of automatic lighting in the home, yard lighting, security warning/recording devices, fences, locks on vehicles and equipment, and by motion detectors. You may choose to use fences to discourage strangers from entering your property.

Neighbors

To minimize problems, maintain good relationships with your neighbors, install protective measures, and cooperate with local authorities. Locate large garage and shop doors so building contents are not visible from public roads. Install powered doors where possible because they are more apt to be kept closed than hand-operated doors. Avoid an abandoned look by use of yard lights, parked cars, noisy dogs, and building lighting. Regularly survey the area so thefts are detected and reported quickly. Contact local law enforcement offices to learn about their rural protection programs and security efforts.

Personal Property

Keep an inventory of property, and permanently mark personal property such as hand tools, household items, and lawn and shop equipment. Keep track of your inventory. Maintain accurate records of your belongings and periodically update your records. This will be a good way to alert you to missing or lost items that could go undetected otherwise. Make sure all equipment serial numbers and vehicle license plate numbers are recorded and you keep your insurance coverage up to date. Videotape insured assets, including your intruder-protection measures, to support any insurance claims. Still photos with dates, identifiable backgrounds, and lists of serial and model numbers can be very helpful. Update this information on a regular basis and store your documentation off site in a safe place, such as a bank safety deposit box.

Rural Mailboxes

Authorities usually advise against placing your name on your mailbox for security reasons, but you should place your street or road identity along with the number that identifies your home. Your name on the mailbox allows would-be intruders to look up your phone number to ensure your home is vacant before they visit. See the section about rural mailboxes and rural mail delivery in Chapter 3, *Utilities and Electrical Services.*

Vacations

Stop deliveries or have a trusted neighbor pick up your mail and newspapers while you are away. Make arrangements to have your driveway scraped in winter, or ask a neighbor to drive down your drive after a snowstorm to leave visible tracks. Place lights and radios or the television on timers. Consider a home security system that notifies a central security company office or the police in case a door or window is opened without disarming the system. Inform local law enforcement offices when you will be away for an extended time, and notify them immediately upon your return so that your presence is not mistaken for that of an intruder. Make sure a trusted neighbor is aware of your absence and knows how to reach you if needed. Do not hide keys on your property. Thieves know to check under nearby rocks, over doors, beneath door mats, and other typical hiding places.

Alarm Systems

Alarm systems provide warning of various problem situations such as arrival of visitors, power failure, water system failure, fire, open doors, and other situations. Figure 8-20 shows how an alarm system might be wired. A video surveillance system may consist of cameras, monitors and recorders, and can be triggered by motion sensors. Record sounds as well as images and the date and time of recording. Modern alarms allow you to remotely check on the status of your home security. In case of alarm, they can notify your cell phone, the police or even trigger lights and sirens to discourage intruders. Make sure detectors stay operational in winter with regular snow removal. Maintain a list of emergency phone numbers at each telephone.

Radon

Radon can be a serious issue in some places in the U.S. and should be considered in evaluating and planning for home safety.

Radon is a clear, odorless, tasteless radioactive gas. It comes from the natural breakdown of uranium in soil and rock and gets into air and water. Radon is found throughout the U.S. The level of radon present can be measured with a simple recorder or with test kits available for less than $100 from many hardware and discount stores. It should be checked every five years. The American Lung Association is concerned about radon's effect on people's lungs via inhaled air. Forced air exhaust ventilation around the home foundation with adequate granular material under the basement floor is usually effective in reducing the radon gas concentration if measurements show levels exceed federal safety guidelines. The US Environmental Protection Agency maintains a radon program and has a broad range of materials about radon mitigation in homes.

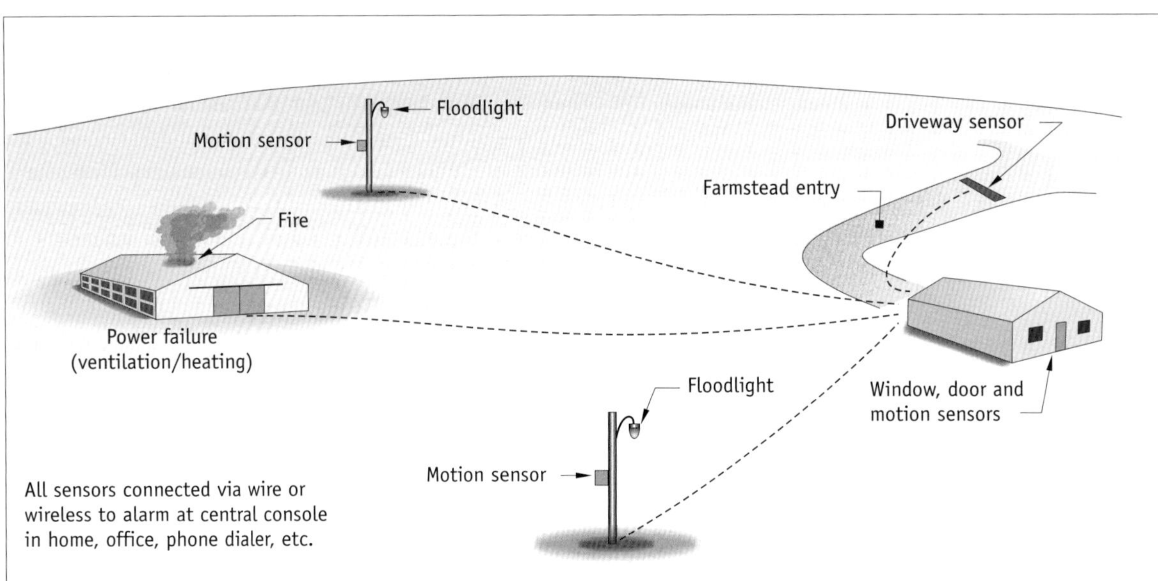

Figure 8-20. Electronic alarm systems can detect various dangers.

Resources

EPA radon web site at
http://www.epa.gov/iaq/radon/index.html

FEMA, http://www.fema.gov,
has some information and publications available for download or purchase from their website, such as:

- *Design Manual for Retrofitting Flood-Prone Residential Structures*
- *Taking Shelter From the Storm: Building a Safe Room Inside Your House*

MidWest Plan Service (MWPS),
http://www.mwps.org

- *Wiring Handbook for Rural Facilities,* MWPS-28, for more information on lightning rod system planning
- *The House Handbook*, MWPS-16, for more information on storm shelters

National Oceanic & Atmosphere Administration (NOAA),
http://www.noaa.gov,
for more information on climate

NEC Article 280—lightning protection

USDA

- Plan No. 5948, *Underground Storm Shelter and Vegetable Storage*
- Plan No. 6376, *Aboveground Storm Shelter*

9

Rural Sanitation and Septic Systems

If you live in a rural area, you are usually responsible for the disposal of any wastes generated at your homestead. This chapter discusses your options for trash disposal and sewage treatment.

Trash Disposal

According to a University of Arizona study (*High Country News*), the average American generates four pounds of household garbage daily, with about 1% of that considered to be toxic waste. Contamination of ground and surface water with nutrients from the break-down of organic matter is a primary concern in its disposal. Toxic materials, present in a number of modern products, including mercury and various manmade organic materials may be introduced into the environment if not disposed of properly. While all of us should minimize the amount of trash we produce, this is especially true in rural areas where there is often no utility service to regularly remove trash or where trash pickup fees are high. There are four common methods of disposing of unwanted material from a family living area, depending on where in the US you live:

1. Burning (where permitted, this is usually limited to wood and wood-based products)
2. Composting
3. Recycling
4. Pickup

Burning

Burning household trash is legal in some locations but not in others. States allowing open burning of household trash generally restrict what may be burned to wood and paper products because toxic gases are given off when some plastics, pressure treated lumber, and other compounds are burned. Some states and counties have specifications for the type of stove or burning container that can be used. Many require a stove where the fire can be controlled (Figure 9-1). Some toxic compounds and harmful particulate matter are even produced from burning leaves. This is largely because open burning or barrel burning takes place at a low temperature. Burning should be avoided if someone in the family has a respiratory illness. Further, there is always a risk of fire spread. Some states have banned all trash burning because of these dangers. If you are considering burning as a means of disposing of household trash, first check with local and state environmental authorities to make sure it is acceptable in your area. You may need to obtain a burning permit from a local fire marshal. If you decide to proceed, look for a burn container that ensures the fire will be contained. Burn only paper or untreated wood products. Lastly, remember that incineration only reduces the volume of waste; it does not eliminate the need for trash disposal since the ashes must still be disposed of. If burning is allowed in your area, follow the rules below:

- Obtain a burning permit if require.
- Check for burning alerts.
- Never leave a fire unattended.
- Always make sure a fire is completely out before leaving the site, or before dark.
- Do not build a fire unless the wind is still and extinguish the fire if the wind increases.
- Have water and a rake or shovel on hand.
- Keep your fire small enough so you can control it.
- Keep all combustible materials at least 10 feet away from a fire.

An out-of-control fire can be extremely dangerous, especially in an isolated area where combustible material such as dry grass

Figure 9-1. Open fire with wire fence and lids—an attempt to control fire spread.

is present. Without help readily available, an out-of-control fire can be very difficult to bring under control.

Plan the location of any incinerator or burn pile carefully. For example, avoid low or wet areas because wet soil will not provide a good base. Locate the incinerator on a stable foundation and where smoke does not blow in the direction of nearby homes or other inhabited buildings and where a spill or blowing ash would not reach storage areas for chemicals, fuel, or fertilizer or any other combustible material, including dry grass or other vegetation. Provide a cover that restricts air-blown ash and embers from leaving the incinerator.

Composting

Most landfills in the U.S. do not accept yard waste. One of the best and most environmentally benign methods of handling yard and garden waste is by composting to kill pathogens and stabilize organics and then recycling the composted material as a soil amendment. A principle benefit of composting is the significant reduction in volume and the generation of a rich organic soil. Composting, like incineration, can significantly reduce the amount of waste that must be disposed of off-farm. Some materials such as steel and plastics are not degradable so should never be composted or burned.

In recent years, the use of homeowner composting has expanded to include organic kitchen wastes in year-round operations. In cold climates, you may wish to insulate or even heat the compost to aid digestion during very cold weather. Well-composted material can be stored easily for use in the spring when it can be used on gardens and lawns. Some homeowners seed their composting material with earthworms, which can speed the digestion process. Cutting larger yard waste into smaller pieces will help speed the process as well. If you add food scraps to your compost, the compost process should be constructed to exclude wildlife pests (raccoons, opossums, and rodents, for example).

As with incineration, siting is important. There may be a small amount of odor at times, so do not locate a composting pile too close to the home. Do not place the composting operation on low ground because it is important to control the amount of water added to the compost process. In some states, it may be necessary to obtain a permit before beginning a household composting operation.

Recycling

While usually more convenient in urban areas, recycling is a good practice almost anywhere you live. Recycling is even more important in rural areas because trash disposal is usually less frequent and more costly than in urban areas (Figure 9-2). Composting or burning of the organic trash stream, removing as many recyclables as possible, and care in shopping can make a very significant reduction in the amount of trash that must be landfilled. The U.S. EPA has documented as much as an 80% reduction in household trash that must be disposed of in land fills when homeowners follow these practices (*High Country News*). Many communities have organizations that periodically accept old containers of paint thinner, used motor oil, partly full pesticide containers, and other toxic materials from homeowners.

Trash Pick-up

The availability, cost, and frequency of rural trash pick-up varies greatly around the country. The best advice is to check with your neighbors to see if this service is avail-

Figure 9-2. Rural recycling station

able in your area. In most rural areas, there will be no public trash pick-up at your home. If there is a trash pickup utility, expect the cost to be considerably higher than rates charged in town because pick-up costs are largely determined by housing density and by distance to the landfill.

Trash transfer sub-stations are available in some areas (Figure 9-3). These are locations where homeowners drop off their trash, usually with a "tipping" fee to help cover operation of the station. Trucks periodically remove the trash from the sub-station and transport it to a landfill. The location of sub-stations in a rural area is usually not as contentious as in urban areas because sub-stations can be located further from populated areas and because there are often few other choices for handling trash. Since federal regulations (Resource Conservation and Recovery Act of 1976, subtitle D) now require expensive lined and monitored landfills, many small-town landfills around the U.S. have closed.

Figure 9-3. Rural Trash transfer sub-station.

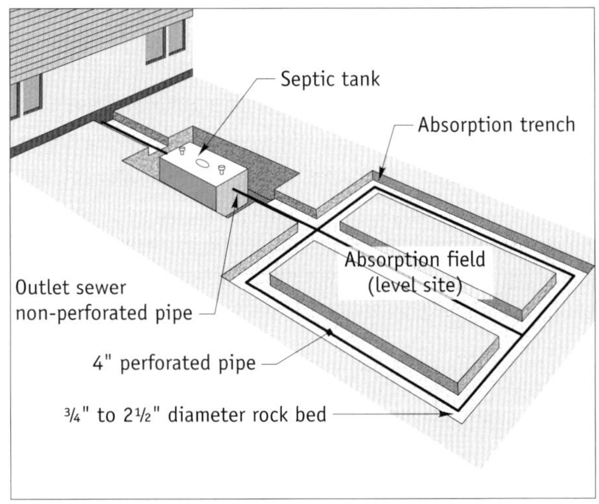

Figure 9-4. Typical layout of a residential onsite system.

Sewage Treatment

If access to a city or municipal sewer system is unavailable, a private onsite sewage treatment and disposal system is needed. Proper design, construction, operation, and maintenance of a wastewater treatment and disposal system will minimize any health hazard to your family, community, or the environment. A permit is normally required before constructing on onsite waste treatment and disposal system. Check your state, county, or town regulations and plumbing codes before starting construction. Get information from your local health department or county extension office. A number of options are available for onsite sewage treatment. Figure 9-4 shows a common layout, and Figure 9-5 shows options for onsite sewage disposal.

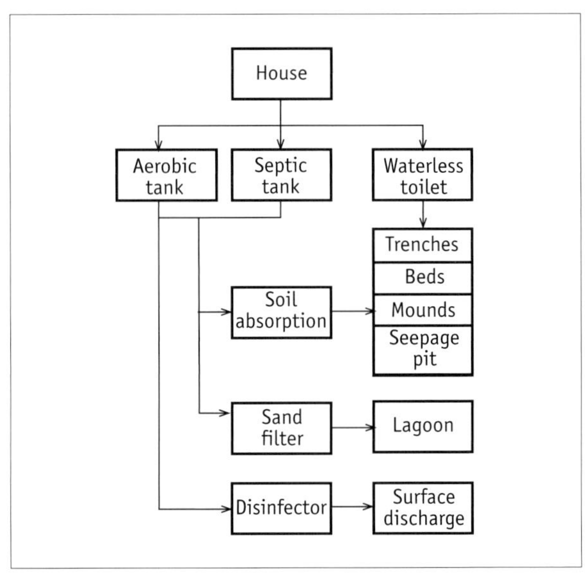

Figure 9-5. Alternative onsite waste treatments.

Onsite Maintenance

If you build your home in a rural area, you probably will have a conventional onsite wastewater system with a septic tank and absorption field. As with any home maintenance, proper upkeep will save you time and money in the long run. Most tanks need to be pumped to clean out accumulated solids (septage) every three to five years, depending on tank size and the amount and quality of solids entering the tank. Have an experienced operator with appropriate equipment clean the septic tank and dispose of the removed material (septage). In some areas, septage pumpers must be licensed. Whatever type of system that gets installed, follow the guidelines for system development and maintenance listed below:

- Pump your tank regularly. Septic tank and drainfield additives advertised to extend the life of septic systems have not been shown to be effective. The cost of having your tank pumped regularly should be considered to be a regular maintenance cost and, when averaged over 36 to 60 months, results in a very low cost per month.
- Keep large vehicles off any part of the septic system before, during, and after construction. In winter, minimize all traffic, including snowmobiles and foot traffic because compaction increases frost penetration. The winter after construction or renovation of a septic system may lead to freezing problems due to bare soil. Cover all bare soil with an 8- to 12-inch layer of straw or hay and weight it down to keep it in place.
- Divert surface runoff water from roofs, patios, driveways, or other paved areas away from the soil absorption field. Grade the area so waterways, drainage ditches, and other water flows, do not drain across the site. Never discharge water from basement footing drains, roof downspouts, or other non-waste water sources into the septic tank or absorption system. Since a septic system is a biological treatment system, minimize the amounts of cleaner, disinfectant, or other toxic materials added to the system.

Selection Criteria for Onsite Systems

Several types of onsite sewage treatment systems can be considered if you need to install a new septic system. You must select a septic system that functions well in your location and is acceptable to local health authorities. The following sections provide an overview of your options. To select the right system, carefully analyze the site. Table 9-1 lists available systems and the site-criteria you must consider. An *x* indicates that a system is feasible.

When a waste system needs maintenance or when property ownership changes, an accurate sketch of the waste system is important. The plan should show the following components:

- Septic tank size and location.
- Openings for pumping out sludge.
- Soil treatment area size and location.
- Direction and distance of the septic tank and treatment area from the house, well, and property line.
- Installer's name and address.

Septic Tanks

Septic tank location with respect to the home is usually determined by the location of suitable soil, the slope of the land, and the location of major bathroom and kitchen plumbing. Locate septic tanks away from your well, at least 10 feet from foundation walls, and straight out from the point where the house sewer pipe goes through the wall. Check local codes for distance requirements between septic tanks, foundation walls, and property lines. Locate the septic tank away from vehicle traffic areas, because excessive loads can damage the tank. Avoid areas subject to flooding or near roof gutter downspouts. Locate the tank convenient to an accessible road so a tank pumping truck can clean the tank. Consider future home improvements such a sidewalks, patios, garages, and storage buildings when selecting the tank site.

The design of soil absorption systems to treat and dispose of household wastewater is largely based on the concept of soil percolation rate and adequate distance to groundwater and/or bedrock. The percolation

Table 9-1. General system selection chart.
This table should be used as an approximation only. Check with your local county health department for restrictions in your area.

	System			
Site criteria	Conventional Trench	Conventional Bed	Mound	Treatment with surface discharge
Soil Percolation rate (minutes per inch)				
1-30	x	x	x	
30-60	x	x	x	
60-120	*		x	
More than 120			x	x
Depth to high water table or creviced bedrock				
More than 10'	x	x	x	
5' - 10'	x	x	x	
2' - 5'	x	x	x	
Less than 2'				x

rate is a measure of the soil's ability to absorb water. Conventional systems (Figure 9-6) are used in most sites and slopes—other than sites with steep slopes and shallow soil to the impermeable layer. Mound systems are more expensive but can overcome some of the limitations of shallow soil depth. See Table 9-1 for specific criteria for selecting a system.

Installation

Septic tanks (Figure 9-7) should be installed as shallow as possible. Some areas in cold climates require 2 inches of polystyrene insulation on the septic tank lid to prevent freezing problems. Septic tank depth depends on house plumbing and whether there is gravity flow from a basement sewer drain. Some states require a basement sump for wastewater generated in the basement if the rest of the system can operate by gravity. Gravity flow from the house sewer to the absorption area, or drainfield, is the most desirable design. Some locations require an effluent lift pump located in a pump chamber downstream from the septic tank to transport wastewater effluent to the absorption field. Sewage lift pumps also can be used if necessary to move wastes from basement bathrooms up to the septic tank. Tanks should have at least two riser pipes that extend to the soil surface. One should be an inspection port over the outlet baffle. The other should be an access port to the tank for pumping. Some areas even require a third riser to provide an inspection port over the inlet baffle. Grade the soil surface away from the risers to divert surface water. Place a secure cover on risers to prevent children from removing it.

Sizing

Check state and local codes for required septic tank size. The minimum septic tank size for any installation is 750 gallons. How much larger a tank you choose will depend on the size of your household. Table 9-2 shows typical tank capacities for residential septic tanks. If you must use a garbage disposal, add at least 50% to the recommended tank size because of the extra load to the system. A better alternative for food waste disposal in rural areas is composting.

Septic tanks can have one or two compartments. Two-compartment septic tanks, or two single-compartment tanks in series, perform better than one, single-compartment tank. The first compartment should

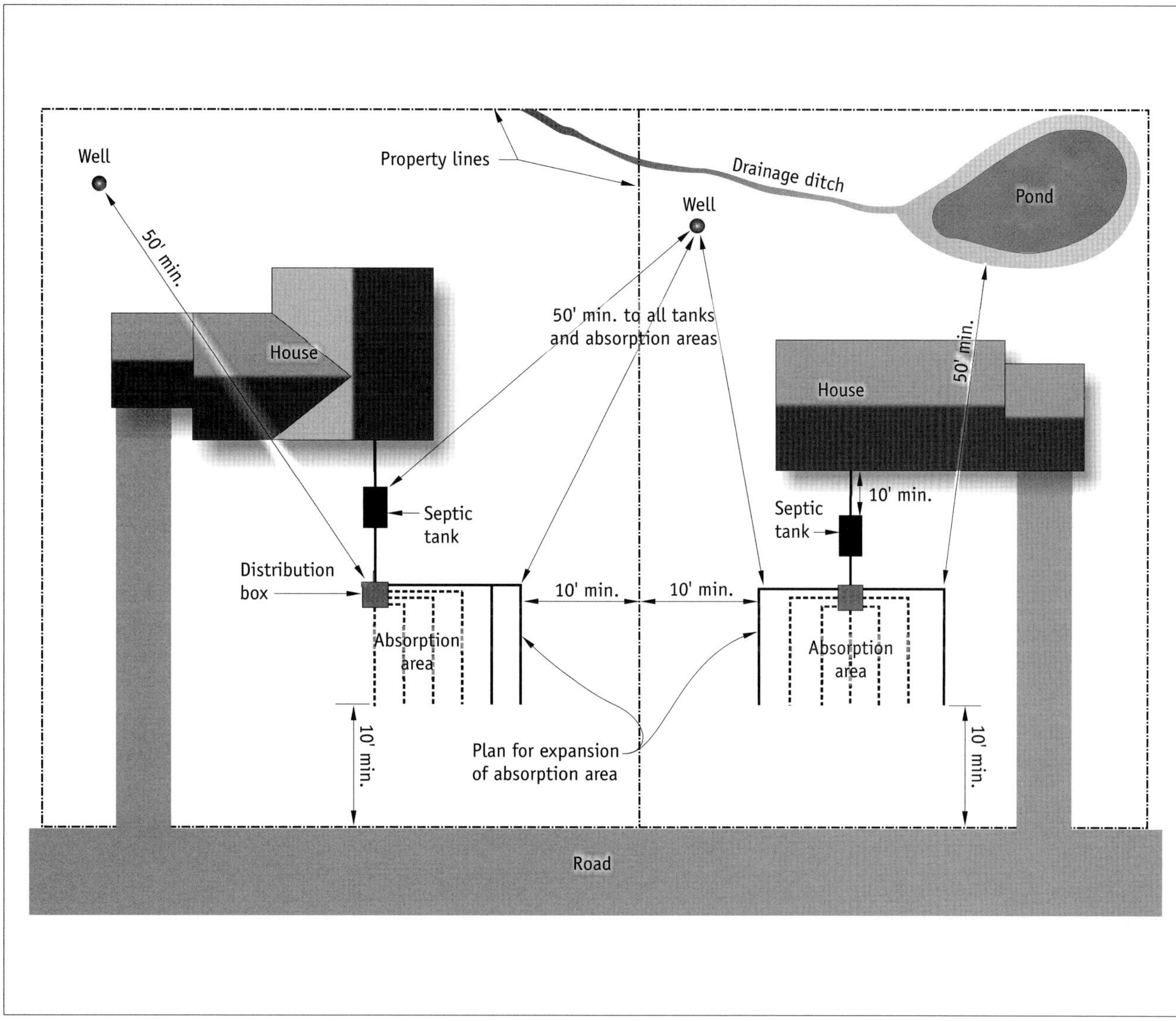

Figure 9-6. Site plan showing septic system location and separation distances.

Table 9-2. Typical liquid capacities for residential septic tanks.
Increase capacity by 50% for kitchen sink garbage disposal. About 150 gal/day (gpd) of sewage/bedroom is assumed.

Number of bedrooms	Minimum liquid Capacity, gal.
2 or less	750
3	1,000
4	1,250
5	1,500

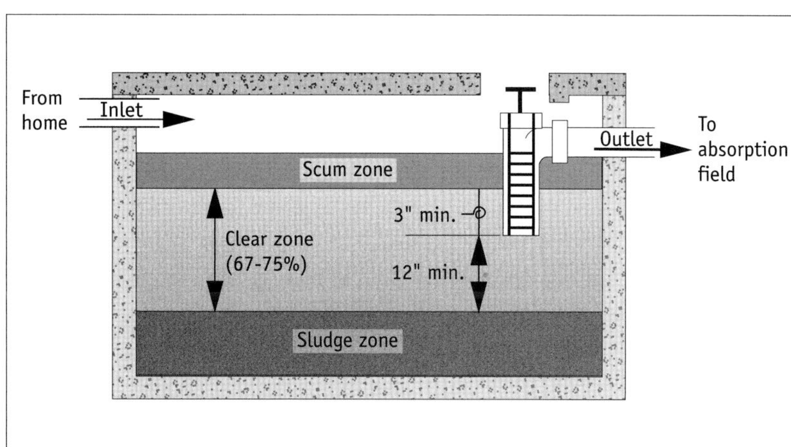

Figure 9-7. Septic tank cross section.

have a capacity of at least 500 gallons (1/2 to 2/3 of the total capacity) to minimize short circuiting and solids carryover to the disposal field. Use proper baffling or effluent filters to enhance the settling of solids and prevent solids from overflowing.

Maintenance

The best designed and operated septic tank/disposal field eventually fails unless sludge is periodically removed from the septic tank. Inadequate maintenance can cause sewage to back up into the house and solids to overflow to the soil disposal area. When solids clog the soil, the disposal area must be abandoned and a new one constructed.

Most tanks need to be pumped every three to five years, depending on tank size and the amount and quality of solids entering the tank. Have an experienced operator with appropriate equipment clean the septic tank and dispose of the removed material (septage) in an approved manner.

Safety

Septic tank gases are dangerous. Methane, an explosive, and hydrogen sulfide, a poison, are the major hazardous gases that can be released during pumping. Torches or other flames near the septic tank opening can cause an explosion. Never lean into or enter a septic tank, even to save someone else, without proper breathing equipment. You could be poisoned or asphyxiated as well. Call a professional if tank maintenance is required.

Soil Absorption Systems

Select the soil absorption field site before locating the house and any other facility. Check local health or zoning regulations for minimum setback distances. If there are no other guidelines, separate the absorption field at least 10 feet from buildings, trees, and property lines, and 50 to 150 feet from wells and other water sources (Figure 9-3).

Locate the soil absorption field where good grass cover is possible. Keep vehicles off the area before, during, and after construction. In winter, minimize all traffic-including snowmobiles and foot traffic because compaction increases frost penetration.

Divert surface runoff water from roofs, patios, driveways, or other paved areas away from the proposed absorption field area. Grade the area so waterways, drainage ditches, and other water flows do not drain across the site. Never discharge water from basement footing drains, roof downspouts, or other sources into or onto the absorption system.

In most areas, the county health department, or similar agency, is responsible for approving and permitting onsite wastewater treatment systems for residences. Professional septic system installers are normally available locally to assist in evaluating prospective sites for percolation rate, depth to creviced bedrock, depth to high seasonal water table (saturated soil condition), depth to impermeable soil layer or rock, and slope. Table 9-3 indicates which types of systems are acceptable for different soil conditions.

Absorption trenches

Absorption trenches are the most common and least costly soil absorption system. Figure 9-8 shows a conventional soil absorption system. Note that a replacement area is set aside and protected in case the system fails and has to be replaced at some point. A biological mat develops on the trench bottom and sidewalls that treats the sewage as it leaves the trench. Freezing rarely occurs in a properly constructed and continuously operated absorption trench. Choose a site that sheds rather than collects surface runoff as shown in Figure 9-9.

Sizing

The trench bottom area required depends on the amount of sewage entering the system daily and soil porosity as determined by the texture and structure of the soil. Soil trenches are normally designed to handle 150 gpd (gallons per day) from each bedroom in the home. In most states, residential trench design is based on the number of bedrooms, rather than the amount of wastewater actually produced in the home because the number of

Table 9-3. General soil evaluation guidelines for conventional and mound systems.
Depths are the minimum required and are measured from the soil surface. "High water table" means high seasonal saturated soil level. Check your local and state Health Office of requirements in your area.

	Units	Conventional Trench	Bed	Mounds
Percolation rate	min/in	1-60	1-60	1-120
Depth to creviced bedrock	ft	3	3	2
Depth to highwater table	ft	3	3	2
Depth to impermeable soil rock	ft	3	3	2-5[a]
Maximum slope	%	24	6	12[b]

a Depends on soil permeability, frost depth, and system layout. Provide enough area to move effluent away. See section on mounds.
b For percolation rates of 1-29 min/in. maximum slope is 12%; for 30-120 min/in, 6%

people in the home may change over time. Table 9-4 shows typical trench loading rates and bottom areas required. Be sure to check the regulations with your local authorities as these values vary from state to state. Table 9-4 uses an estimated wastewater production of 150 gallons per day per bedroom. If the soil percolation rate is outside the range shown, the soil is unsuitable for standard soil treatment units (refer to section on mounds and alternative systems).

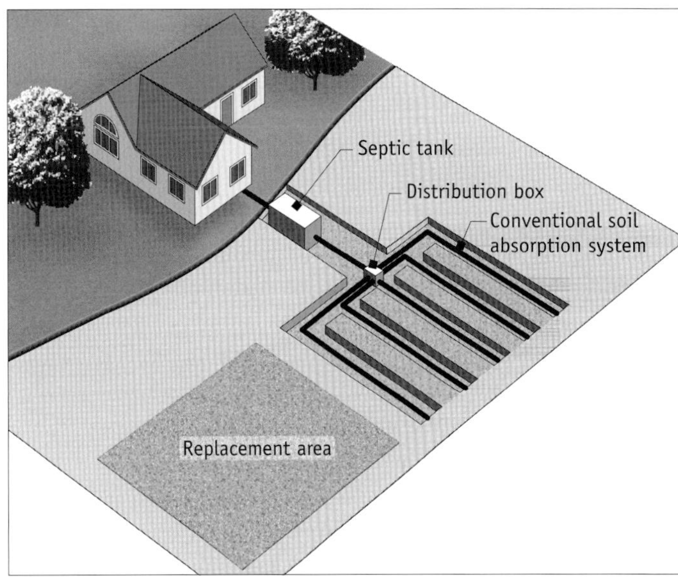

Figure 9-8. Soil absorption system.

Table 9-4. Typical trench loading rates and bottom areas required.

Percolation rate	Trench loading rate Gal/Sq.ft.-day
1 to 5 1.2	
6 to 150.8	
16 to 300.6	
31 to 450.5	
46 to 600.45	
	ft²/bedrm
1 to 5125	
6 to 15190	
16 to 30250	
31 to 45300	
46 to 60330	

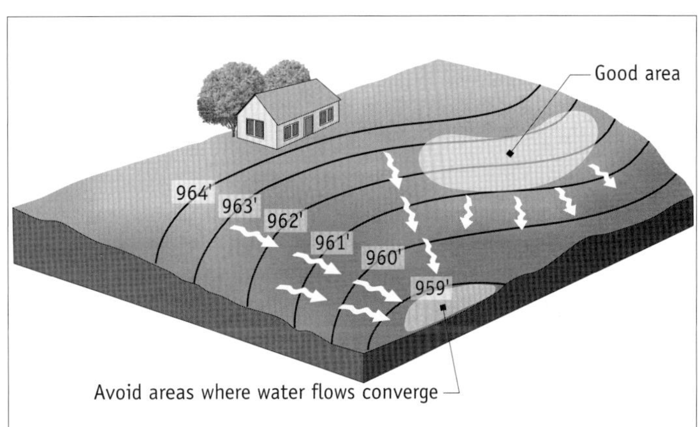

Figure 9-9. Select a convex rather than a concave site which can dispense applied effluent.
Source: Purdue University

Rock around and below the distribution pipe distributes the effluent more uniformly over the trench bottom and provides a porous storage zone until the liquid can infiltrate into the soil.

Place the bottom of the trench 18 to 36 inches below the soil surface. A typical trench width is 36 inches. Limit single trenches to 100 feet long. Use at least two trenches, unless the needed trench length is less than 50 feet. Leave at least 5 feet of undisturbed soil between trenches. The following example illustrates how an absorption area is designed.

Example

Find the size of soil absorption area required for a three-bedroom home.

1. Determine the amount of wastewater requiring disposal. 3 bedrooms x 150 gpd/bedroom = 450 gpd.
2. Determine the soil percolation rate. This normally requires soil borings at the proposed site. For this example, assume the percolation rate is 35 minutes per inch (MPI).
3. Find the loading rate for 35 MPI soil. From Table 9-4, it's 0.5 gal/ft^2 per day.
4. To determine the trench bottom area required, divide the estimated flow rate (Step. 1) by the soil loading rate (Step 3). 450 gpd/0.5 gal/ft^2 – day = 900 ft^2.
5. Another way to find bottom area is to use the lower half of Table 9-4. For 35 MPI, the trench bottom area is 300 ft^2 per bedroom. Multiply the area by the number of bedrooms. 300 ft^2/bedroom x 3 bedrooms = 900 ft^2.
6. Find the trench length required for the desired width. Assuming a 36" (3') width trench; total length is the required trench bottom area divided by the width in feet. 900 ft^2 ÷ 3' = 300 lineal feet.
7. Single trench length is normally limited to a maximum of 100', so divide the 300' into three or more trenches.

Considerations in constructing an absorption field

- Soil conditions and construction practices affect how well the absorption system functions. Construction when the soil is too wet can cause system failure in a short time.
- If the soil crumbles when you try to roll it into a ribbon, it is dry enough for construction, and there is less danger of compacting or smearing the soil. Either of these can damage the soil's ability to absorb wastewater effluent. This is especially a problem in soils with high clay content.
- Absorption trenches are normally excavated with a backhoe. Depth can vary from 18 to 48 inches but are typically 24 to 30 inches. A 30-inch depth allows for 12 inches of rock below the 4-inch distribution pipe and 2 inches of rock cover plus 12 inches of soil over the pipe. The bottom of the trench should be level its entire length with no slope or grade.
- Use rock that is clean and free of dust, dirt, and debris. It should be at least ¾-inch, but no greater than 2 ½ inches in diameter. Do not use pea gravel or pit run gravel, and use caution when selecting crushed limestone. Calcite limestone is soft, deteriorates in contact with septic tank effluent, and clogs the soil.
- Distribution pipes are usually 4-inch rigid plastic. Install the distribution pipe on a uniform grade with a maximum of a 2-inch drop in 100 feet, but make sure to comply with local and state codes.
- Cover the top of the distribution pipe with at least 2 inches of clean rock. Cover the top of the rock with synthetic geotextile fabric or other permeable material. This helps prevent the rock from becoming clogged by the soil backfill.
- Backfill over the rock with soil of the same texture as the original soil. Crown the backfill 4 to 6 inches above finished grade to allow for settling and to shed rain and runoff (Figure 9-10).

- Establish a grass cover over the absorption area as soon as possible. If grass cover is not established before winter, cover the trenches with straw or hay to prevent freezing. On sites with a high seasonal water table, use trenches up to 100 feet long, and space them as far apart as practical. Trenches can extend 200 feet along a slope if a drop box supplies effluent at the center, so each trench is no more than 100 feet long. Put a 4-inch observation pipe, extending to the trench bottom, at the end of each trench. If water is ponding at the end of a trench, take the trench out of service for at least six months by plugging the distribution pipe at the distribution box (Figure 9-11).

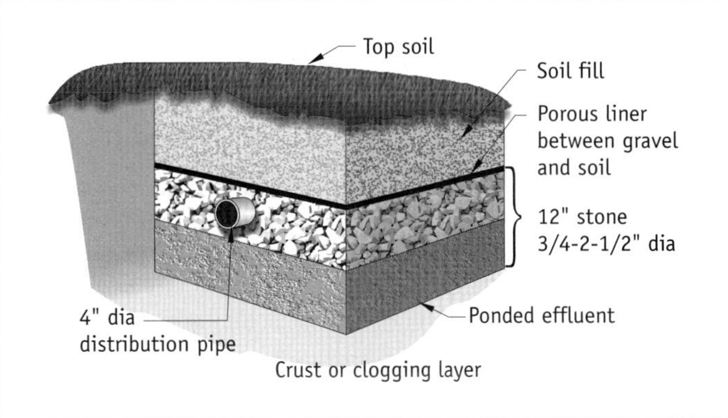

Figure 9-10. Cross-section of typical trench system.

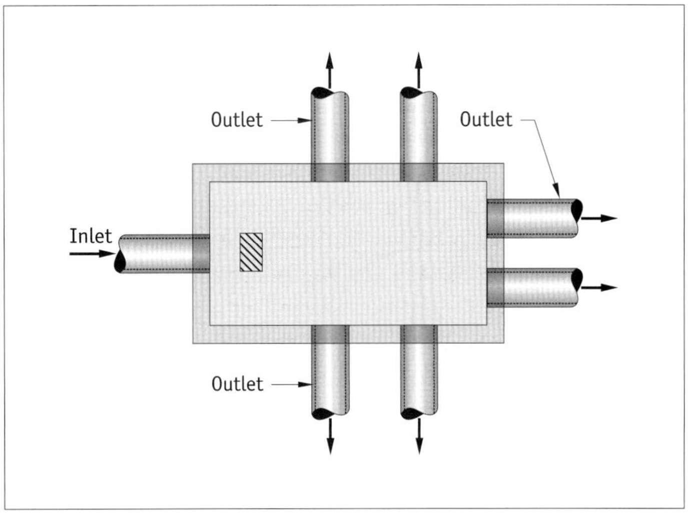

Fig 9-11. Distribution box or D-box that distributes septic tank effluent to the soil absorption trenches, plan view.

Elevated Sand-mound Systems

Use mounds for onsite wastewater disposal on sites that have any of the following soils:

- Slowly permeable soils with or without high water tables.
- Shallow permeable soils with creviced or porous bedrock.
- Permeable soils with high water tables.

A septic tank and mound system includes a septic tank, pumping chamber, and mound (Figure 9-12). A pump elevates effluent to the mound and pressurizes the distribution system. The mound has fill material, an

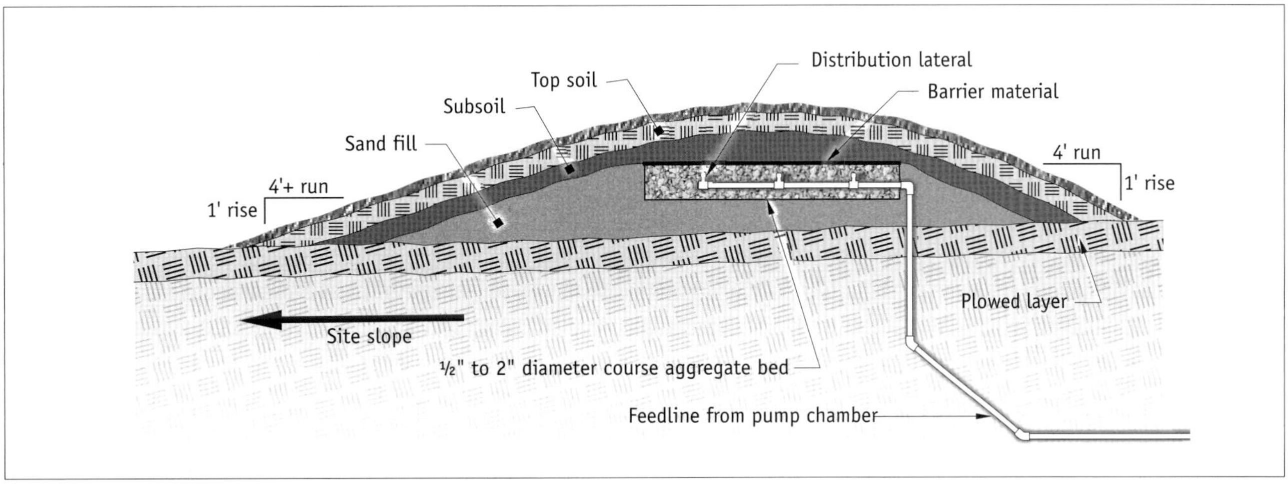

Figure 9-12. Cross-section of elevated sand mound built on original ground surface.

absorption area, a distribution system, a soil cap to shape the top surface, and topsoil. Effluent is pumped into the absorption area through the distribution system. Fill material pre-treats the effluent before it passes into the natural soil. The topsoil provides frost protection, retains moisture for vegetation, and promotes runoff of precipitation.

Soil and site requirements

Table 9-5 summarizes typical soil and site limitations for mound construction. Contact your local health department for restrictions that apply in your area.

Generally, sites with large trees, numerous smaller trees, or large boulders are unsuitable for a mound system because of difficulty in preparing the surface and the reduced infiltration area beneath the mound. Rock fragments, tree roots, stumps, and boulders reduce the amount of soil available for proper purification.

The recommended separation distance between the toe of a mound (where the mound meets the original soil level) and buildings, property lines, and driveways is 10 feet for 3:1 side slope mounds. On slowly permeable soils, a separation distance of 30 feet is recommended. More clearance also permits mound repair if seepage occurs.

Fill material

The mound requires a "medium" sand texture fill beneath and beside the absorption area (trenches or bed). The sand need not be screened or washed, but it should not have more than 10% fines.

The topsoil over the mound and a medium or medium-fine textured soil cap provide higher water holding capacity for plant growth and minimize rainfall infiltration. Often, excavated soil from the original site can be re-used. Place good topsoil 6 inches deep over the entire mound to promote good vegetation.

Absorption bed in a mound

Water table height and soil permeability determine absorption bed shape. Use a long absorption trench 4 to 6 feet wide on sites with a high water table and slowly permeable soils. For permeable soils, a rectangular bed up to 10 feet wide can be used. Use either layout for shallow permeable soils over creviced or porous bedrock. Place at least 12 inches of rock under and 2 inches above the distribution pipe lateral. Use ¾- to 2½-inch washed rock or crushed stone. Do not use soft limestone. Orient the absorption bed along the contour so the effluent spreads out as it moves down slope.

The absorption rate for a medium sand fill is 1.2 gal/ft^2-day. For a three-bedroom home and medium sand, the required absorption area is 3 bedrooms x 150 gpd/bedroom ÷ 1.2 gal/ ft^2-day = 375 ft^2.

Pressure distribution system in a mound

In a pressure distribution system, the wastewater effluent leaving the septic tank is applied uniformly over the entire absorption area. This is achieved by using a pump to provide pressure and small diameter pipes to distribute a small amount of effluent at regular intervals during the day. The pipe laterals and manifolds must drain after every dosing (addition of effluent).

Pressure Distribution Trench Systems

The pump and pipes in a pressure distribution trench system are similar to those used in elevated sand mounds. Trench systems are less expensive to construct than mounds but more expensive than a conventional gravity system. Trench systems are used when gravity cannot be used to convey septic tank effluent to the soil disposal field such as when the disposal field area is at a higher elevation than the septic tank discharge (Figure 9-13). Pressure distribution with a pump provides more even distribution and therefore more uniform loading over the soil disposal area than does a gravity-loaded system.

A general recommendation is to dose four times a day in sandy soils and in mounds, with larger doses added less frequently in heavier soils. Distribution laterals should drain after each dose.

A pressure system includes a pumping chamber, pump, and alarm system

Table 9-5. Soil and site restrictions on mound systems.

	Soil group		
Restricting factor	**Slowly permeable soils**	**Permeable soils with porous bedrock**	**Permeable with high water tables**
Percolation rate, MPI	60-120	1-60	1-60
Test depth[a]	24"	16"	16"
Depth to porous rock	24"	24"	24"
Depth to high water tables	24"	24"	24"
Depth to impermeable soil layer or rock strata	2'-5'[b.]	2'-5'	2'-5'[b.]
Depth to 50% by volume rock fragments:	24"	24"	24"
Slope	6%	12%[c]	12%[c]

a Percolation test depth, unless there is a more restrictive horizon above. If perched water is at 24", hold test depth to 16".
b See discussion in text. Depth depends on soil permeability, length, and climate.
c For percolation rates of 1-29 min/in, maximum slope is 12%; for 30-60 min/in, 6%.

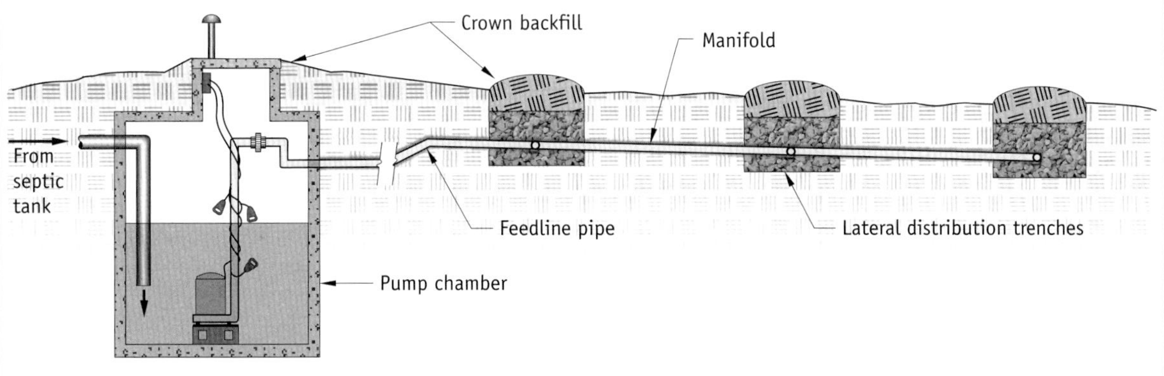

Figure 9-13. Pressure distribution used to load a trench.

(Figure 9-14). Size the pumping chamber to provide the desired dosing volume, space for the controls and pump, and extra volume for a malfunction and pipe draining after the pump shuts off.

The pumping chamber and septic tank must be watertight, so groundwater is not pumped to the absorption system.

Vent the pumping chamber for proper pump performance. If possible, mount all electrical controls outside the tank and protect against lightening damage. Provide an easy pump disconnect, such as a union, for pump removal in case of failure.

To prolong pump life, keep the pump submerged for cooling.

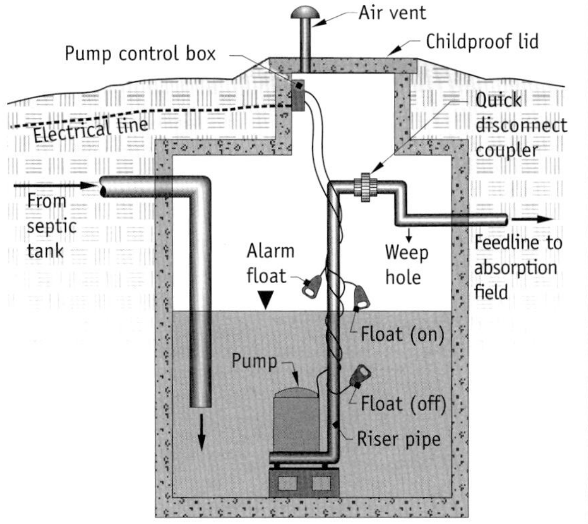

Figure 9-14. Cross section of a typical pumping chamber.

Pre-Treatment Systems

Several systems are available to pre-treat the septic tank effluent before it reaches the soil infiltration area. Partially treating the wastewater removes some of the biological load from the soil and usually improves the performance of the system.

Aeration systems

A septic tank treats wastes by bacteria which live in an anaerobic (without air or oxygen) environment; while an aeration system treats waste with aerobic bacteria (requires air). An aeration system treats waste more completely than a septic tank, but the effluent is still not of sufficient quality to be discharged directly on the soil surface or to a waterway in most states. An aeration system has higher operating and maintenance costs than conventional trench systems. If possible, a service agreement should be purchased with an aerobic unit. If effluent is discharged to a soil absorption area, design the absorption field the same size as for a conventional system.

In most states, treated effluent from an aeration system must be disposed of in a soil absorption system, although some state regulations allow a downsizing of the field if the wastewater is pretreated. In a few areas, pretreated wastewater can be discharged, but additional treatment or disinfection is usually required. Check with your local environmental regulatory authority for restrictions in your area.

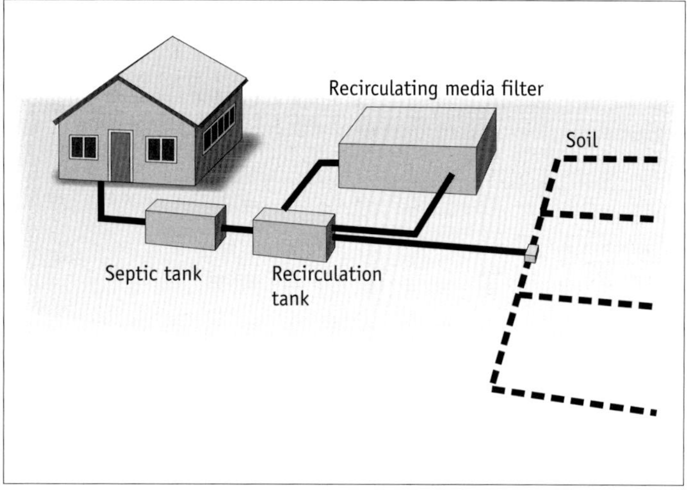

Figure 9-15. Recirculating sand filter treats effluent before it reaches the absorption trenches.

Recirculating sand filters

Where soils have a percolation rate too slow to construct an absorption field or mound, a recirculating sand filter (Figure 9-15) can be used to treat septic tank effluent before it is discharged to the soil system. Both one-pass and recirculating sand filters are used.

In a recirculating sand filter, wastewater passes from a sump at the end of the septic tank through the filter several times in order to treat the wastewater. The primary advantage of recirculating sand filters is the nitrogen content of the wastewater effluent can be markedly reduced. This is important in sandy soils where septic tank effluent could reach the groundwater.

Constructed Wetlands

A constructed wetland can be an economical and effective method of treating septic tank effluent (Figure 9-16). They are effective year around, even in cold climates.

Community Cluster Systems

A cluster system with a common soil treatment unit may be a satisfactory waste treatment system for two or more property owners (Figure 9-17). Each home typically has a septic tank. The effluent usually flows by gravity into a collector line to a pumping station, where the wastewater is treated and disposed of in a large soil absorption system. If adequate slope is not available, intermediate pump stations may be needed along the collector line or at each home. In some cases, it may be necessary to install a pretreatment system such as a recirculating sand filter, wetland, or aeration tank before the effluent goes to the soil absorption system.

Property owners must agree on the organizational, operational, and maintenance details before the system can be designed for the system to be successful. In some states, there is grant money available to assist with design and construction of cluster systems. Such a system requires the community to work closely with professional soil scientists, engineers, professional maintainers and the local regulatory authority.

Holding Tanks

For a residence on a small lot with no suitable soil treatment area and in an area too isolated for a collector system, a holding tank will be needed. Sewage is discharged to a holding tank and later pumped and transported to a treatment and disposal system.

Holding tanks are usually allowed only for an existing residence and not for a new home. Check state and local regulations to determine if they are allowed in your area.

The high cost of hauling sewage makes the holding tank undesirable. The cost can be minimized to a degree by conserving water.

Locate the holding tank so it is accessible to a pump truck under all weather conditions. Check local regulations for required separation distances from wells, water pipes, buildings, and property lines.

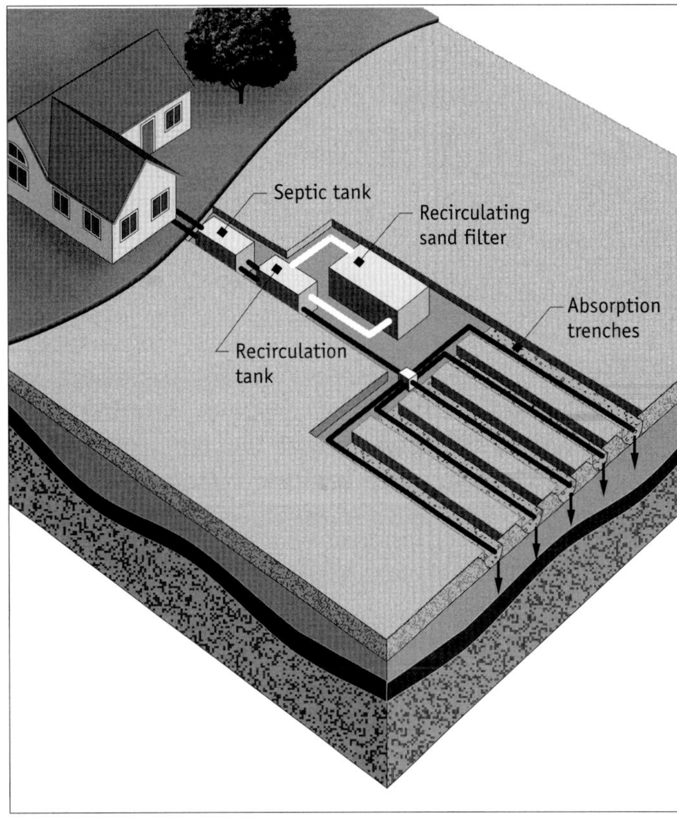

Figure 9-16. Constructed wetland used to pre-treat septic tank effluent before it reaches the soil.

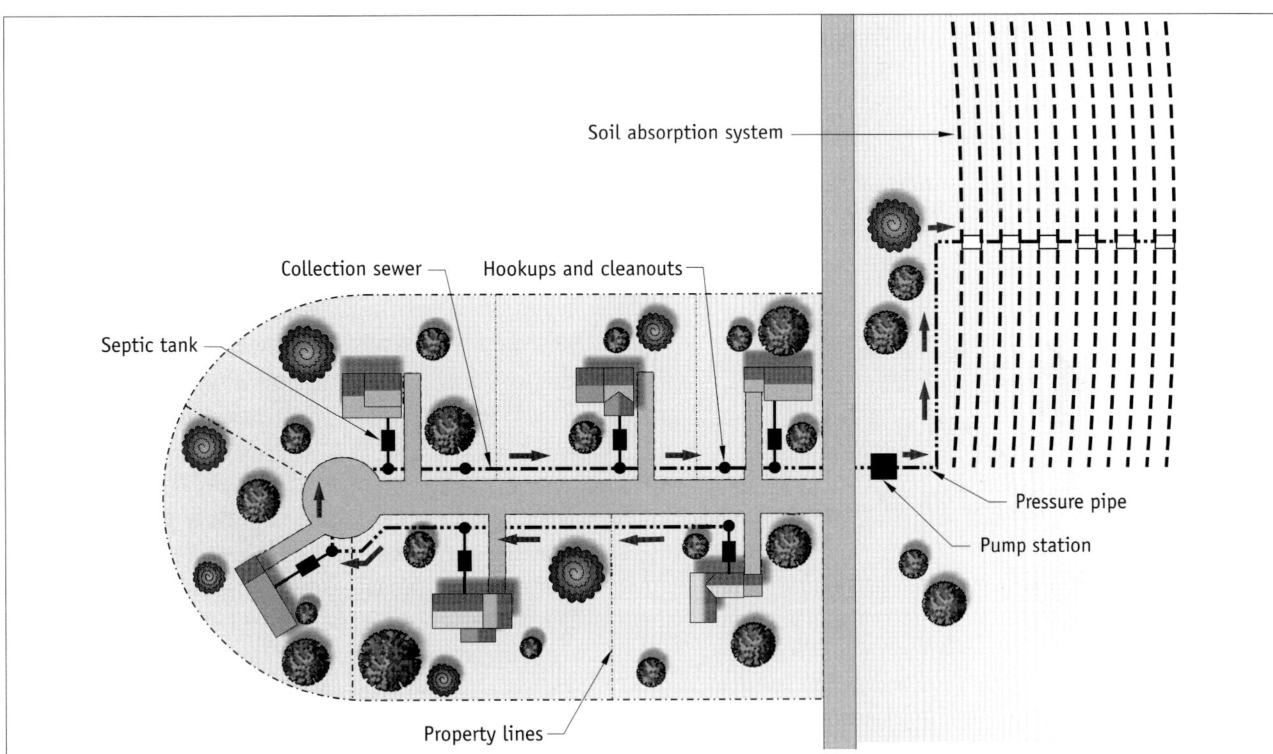

Figure 9-17. Collector sewer system with common soil treatment area.

Install a water meter in your water system beyond the pressure tank and outside spigots to record all liquid wastes flowing into the holding tank. The charge for hauling the sewage should correspond with the water meter readings. If not, search for and fix leaks that are allowing ground water to enter the system. It can also help you evaluate your water conservation efforts.

Resources

Composting to Reduce the Waste Stream: A Guide to Small Scale Food and Yard Waster Composting, NRAES-43, Northeast Regional Agricultural Engineering Service, 152 Riley-Robb Hall, Ithaca, NY, 14853

The Biocycle Guide to Yard Waste Composting. 1989. Staff of *Biocycle*, the Journal Of Waste Recycling. The JG Press, Inc. Emmaus, Pennsylvania.

EPA, *Onsite Wastewater Treatment Manual*, EPA/625/R-00/008, EPA National Risk Management Laboratory, Cincinnati, OH 45268. Manual online at http://www.epa.gov/epahome/publications.htm and do a search using the document number above.

High Country News. March 7, 1994. http://hcn.org

MidWest Plan Service (MWPS), http://www.mwps.org, 122 Davidson Hall, Iowa State University, Ames, Iowa 50011-3080

- *Residential Onsite Wastewater Treatment Systems: An Operation and Maintenance Service Provider Program*, CIDWT. 2006. Consortium of Institutes for Decentralized Wastewater Treatment.
- *On-site Domestic Sewage Disposal Handbook*, MWPS-24. 1982.

10

Maintenance Scheduling Calendar

Maintenance calendars protect the investment in your home while spreading the work over the year. It depends on your location, type of home, etc. Most items can be checked at any time of the year; a regular schedule helps to do it. It is also helpful to keep a notebook listing contact information for all utility providers: phone, electric, gas, septic system, etc. Most apply no matter where you live.

Example

Month	Maintenance
January	■ Check LP gas supply.
February	■ Check LP gas supply.
March	■ Check LP gas supply. ■ Test well water for nitrates & bacterial contamination. ■ Check lightning protection grounding on home and out-buildings.
April	■ Check air conditioner (remove cover, filters, operation, clear leaves and debris and keep shrubs, grass & weeds away for free air flow). ■ Refrigerator and freezer maintenance (clean coils & drainage tray). ■ Prepare garden area. ■ Remove storm windows and windows as needed. ■ Check lawn/garden/animal water supplies. ■ Contract LP gas for fall.
May	■ Clean roof gutters and downspouts of debris. ■ Clean windows. ■ Make sure that basement and garage floor drains are functional and clear. ■ Check for odors. ■ Check basement sump pump and controls.
June	■ Sharpened lawn mower blades, chain saw, etc. , check oil levels. ■ Check level of sludge and scum in septic tank—or effluent filter if present. ■ Have tank cleaned and pumped if needed.
July	■ Check and clean fireplace chimneys & wood stoves as needed.
August	■ Check roof decking for soft spots, broken shingles, leaks, etc.
September	■ Check furnace/heater (filters, operation), check LP gas supply. ■ Compost leaves and garden waste.
October	■ Winterize lawn mower and other gas powered motors. ■ Install storm windows and windows as needed. ■ Drain any water lines or fixtures that are not frost-proof. ■ Drain and store lawn and garden hoses. ■ Clean roof gutters and downspouts of debris. ■ Prepare snow moving equipment.
November	■ Check LP gas supply.
December	■ Replace batteries in smoke and carbon monoxide detectors. ■ Check that fire extinguishers and fire exit ladders are in their proper location.

Multi-month — Multi-year checks:

- Every other year, test water supply for iron, manganese, chloride, fluoride, pH and hardness.
- Every other year have service person check furnace and air conditioner.
- Fertilize lawn every other month or according to local recommendations during the growing season.
- Every 3-5 years—clean and pump septic tank; check lowest level of home for radon.

Your Maintenance Schedule

Month	Maintenance
January	
February	
March	
April	
May	
June	
July	
August	
September	
October	
November	
December	

Multi-month — Multi-year checks:

-
-
-
-
-
-
-
-
-

Index